NoMAD

The Autophagy Path to Healing

NoMAD

The Autophagy Path to Healing

Copyright © *Levitas One*, 2024
All Rights Reserved

This book is subject to the condition that no part of this book is to be reproduced, transmitted in any form or means; electronic or mechanical, stored in a retrieval system, photocopied, recorded, scanned, or otherwise. Any of these actions require the proper written permission of the author.

What are the NoMAD Plans?

Developed by Dr Ash Kapoor, the NoMAD Plans represent a transformative approach to health and wellness that combines the wisdom of ancestral practices with contemporary medical insights. The name "NoMAD" not only suggests a journey through the intricate realm of health but also stands for its foundational principles: Nutritional Optimisation, Mindful Adaptation, and Detoxification.

At the heart of NoMAD is the 6 R Framework—Restore, Release, Repair, Renew, Reframe, and Represent. This methodology addresses the root causes of illness, combats chronic inflammation, and cultivates authentic vitality, guiding individuals through a transformative process.

Tailored specifically to each individual, NoMAD journeys are meticulously crafted to rebalance the body, strengthen the mind, and rejuvenate overall health. By integrating ancestral practices with cutting-edge, innovative treatments—all under strict medical oversight—NoMAD Plans offer a personalised pathway to sustainable, long-lasting well-being that resonates with your unique life circumstances.

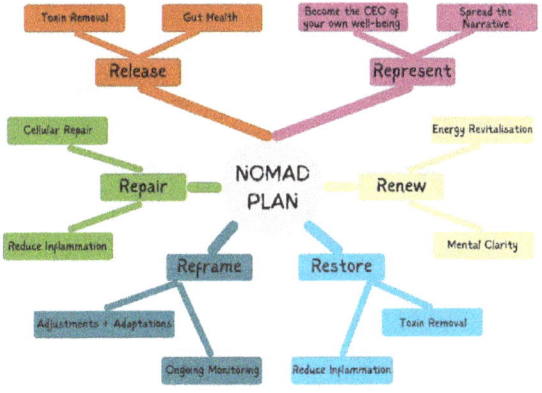

Levitas One:
"As Is In, As Is Out"

Reflecting the belief that our internal well-being is mirrored in our external environment. Founded by Dr. Ash Kapoor, Levitas One serves as the vehicle for delivering NoMAD's treatment plans. It envisions a healthcare future where patients are at the centre of a fully integrated, multidisciplinary approach. Guided by Nomads 6 Rs— Restore, Release, Repair, Renew, Reframe, and Represent— Levitas One empowers self-care through personalised guidance and minimal intervention, promoting long-term health, balance, and sustainability.

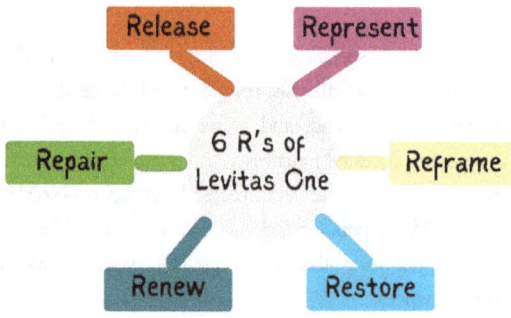

To Alex Millward,

This book is dedicated to you, a key team member from the start. The NoMAD Autophagy path began with humble beginnings at The Mews Practice and has since evolved into guides that are improving hundreds of lives, soon to reach thousands through Levitas One. This could not have been accomplished without your commitment, dedication, passion, and most importantly, the joyfulness that breathed life into NoMAD's inception.

With gratitude and love,

Ash

Contents

Foreword ix

Introduction: 1
   The Journey to Simplicity and Renewal 1

Chapter 1: The Power of Less: A Philosophy of Simplification 5
   The Case for Less 5
   Lessons from Cultural Simplicity 6
   Hiiggi: Spiritual Minimalism 6
   Famous Figures Who Embody Simplicity 7
   Billionaires Who Choose Simplicity 7
   The Benefits of Simplified Living 8
   Embracing the Power of Less 9

Chapter 2: Autophagy Unveiled 10
   2.1 What is Autophagy? 10
   2.2 The History and Evolution of Autophagy 11
   2.3 Mechanisms of Autophagy 11
   2.4 Autophagy and Disease 13
   2.5 Balancing Autophagy and Growth 13
   2.6 The Bigger Picture 14
   2.7 Autophagy: The Key to Thriving 14

Chapter 3: Fasting: A Historical Journey Across Rituals, Cultures, and Science 18
   Fasting in the Ancient World: The Philosophers and Tribes 18
   Hippocrates: The Father of Medicine 18
   Aristotle and Philosophical Discipline 19
   Spartacus and the Warrior Tribes 19
   Hormesis: The Sweet Spot Between Benefit and Risk 19
   Reassurance About Autophagy 20
   The Science of Survival Without Food and Water 20
   How Long Can the Human Body Survive Without Food? 20
   The Transition Stages During Prolonged Fasting 21
   How Long Can the Human Body Survive Without Water? 22
   Fasting as Therapy: Balancing Benefit and Caution 22
   A Timeless Practice for Body and Spirit 22

Chapter 4: Modern Diets and Autophagy 24
   4.1 Overview of Popular Diets 24
   Atkins Diet 24
   Mediterranean Diet 24
   Vegan Diet 25
   Paleo Diet 25
   Ketogenic Diet 25

Intermittent Fasting (IF) 25
Blood Type Diet 25
Warrior Diet 26
Carnivore Diet 26
Whole30 Diet 26
Dash Diet 26
Raw Food Diet 26
Low-FODMAP Diet 27
Anti-Inflammatory Diet 27
Detox Diets 27
4.2 The Role of Diet in Supporting Autophagy 27
Metabolic Plasticity and the Need for Variation 28
4.3 Fasting: The Universal Autophagy Trigger 28
4.4 The Levitas Perspective 29
Conclusion 29

Chapter 5: The Science Behind Autophagy 30
5.1 Cellular Restoration and Autophagy 30
5.2 The Trigger Mechanisms 31
5.3 The Work of Dr. Yoshinori Ohsumi 31
5.4 Autophagy's Role in Disease Prevention 32
5.5 Autophagy and Anti-Aging 32
5.6 Supporting Autophagy in Daily Life 33
Conclusion 34

Chapter 6 The NoMAD Autophagy Plan: Detoxification and Holistic Renewal 36
NoMAD's Phases for Holistic Detoxification 36
Why Illness Can Arise During Fasting 37
How We Mitigate Mineral Loss During a Water Fast 37
Gender-Specific Fasting Variations and Optimal Windows 38
Measuring Success: 39
Glucose-Ketone Index and Ketosis Levels 39
What is the The Added Benefit of NoMAD 40
How NoMAD Helps You Achieve Success 40
How Drips and Nanocelle Technology Support the NoMAD Autophagy Plan 41
1. Drips to Maintain Autophagy 41
2. Drips to Support Rebuilding After Fasting 41
3. Nanocelle Technology for At-Home Use 42
The NoMAD Philosophy: Feeding the Cell, Not the Stomach 43
How Drips, Nanocelle Technology, NAD+, and Glutathione Support the NoMAD Autophagy Plan 43
1. Drips to Maintain Autophagy 44
2. Drips to Support Recovery and Rebuilding After Fasting 44
3. Nanocelle Technology for At-Home Use 45
4. The Value of Glutathione in Detoxification 45
5. The Role of NAD+ in Fasting and Recovery 46

Post-Fasting Phase: Transitioning to Sustained Nutrition 46
Phase 1: Two Meals a Day – Predominantly Vegetables, Protein, and Healthy Fats 47
Phase 2: Reintroducing Low Glycemic Load (GL) Carbohydrates 47
Challenges and Observations 48
The Broad Benefits of the NoMAD Autophagy Plan 48
The 6R Framework and the RELEASE Series 49
1. Brain Detox 49
2. Muscle Detox 49
3. Gut Detox 49
4. Liver and Kidney Detox 50
5. Lymphatic Detox 50
6. Heavy Metal Detox 50
All Systems Working Together 50
The NoMAD Philosophy: A Unified Approach to Health 51

Chapter 7 Real-Life Success Stories 53
7.1 Case Study One: Weight Loss and Energy 53
7.2 Case Study Two: Overcoming ADHD and Insomnia 54
7.3 Case Study Three: Reversing Early-Onset Diabetes 55
7.4 Case Study Four: Managing Cravings and Stress 56

Chapter 8: Decompressing the Body Through Autophagy 59
8.1 Autophagy: The Body's Ultimate Reset 59
8.2 The Gut: Where Health Begins 60
Autophagy in the Gut: 60
83 The Liver: The Body's Detox Hub 61
Autophagy in the Liver: 61
8.4 The Kidneys: Precision Filters 61
Autophagy in the Kidneys: 61
8.5 The Brain: A Clearer Mind 62
Autophagy in the Brain: 62
8.6 The Heart: Resilient and Efficient 62
Autophagy in the Heart: 62
8.7 Muscles: Strength and Recovery 63
Autophagy in Muscles: 63
8.8 The Spleen: Immunity's Gatekeeper 63
8.9 Cancer and Autophagy: Supporting Modern Therapies 64
Centers Using Autophagy in Cancer Care: 64
8.10 Systemic Benefits: Healing from Within 64
Conclusion 65

Chapter 9: Final Thoughts 67

Glossary 70
References 74

Foreword

In a world where society seems to constantly look for quick fixes, it feels to me there is a danger for us to lose touch or belief in our body's natural ability to heal and thrive on its own accord. **Autophagy**—the process of cellular self-cleaning—is an incredibly powerful mechanism that helps achieve true health and longevity, far beyond simply managing symptoms.

Having gone through this transformative process myself under the guidance of the brilliant Dr Kapoor, I've witnessed the benefits firsthand —from increased energy and mental clarity to deeper emotional balance and resilience. It really is super powerful. This book explores the science of autophagy and its profound impact, showing you how to activate it through fasting, nutrition, and mindful living.

From what I experienced I would say Autophagy is the secret to rejuvenating bodies, sharpening minds and uplifting spirits.

Dr Kapoor's NoMAD approach merges modern medical knowledge with ancient wisdom, offering practical, personalised strategies to harness autophagy and unlock your best self. Whether you seek improved health, mental sharpness, or emotional wellbeing, this guide provides a roadmap to renewed vitality and a life of true balance and fulfillment.

— *Phil Spencer*

Introduction:

The Journey to Simplicity and Renewal

In the fast-paced complexity of modern life, we often lose sight of the simplicity that once guided human health and well-being. The constant noise—be it the demands of daily routines, over-processed diets, or information overload—leaves us disconnected from the innate ability of our bodies to heal, restore, and thrive. Yet, beneath the layers of distraction lies a remarkable, natural process within us, one that holds the key to renewal: **autophagy**.

Autophagy, meaning "self-eating," is the body's way of recycling damaged cells, proteins, and other waste to promote cellular repair and regeneration. This process isn't just a biochemical phenomenon—it's the foundation for profound health benefits, from detoxifying the body to improving energy, focus, and longevity. By embracing strategies that enhance autophagy, we allow the body to clear its cellular clutter and re-establish balance, vitality, and resilience. This book is your guide to unlocking the power of autophagy and achieving a life of renewed health and simplicity.

At the heart of this journey is the philosophy that less is often more. In a world of overconsumption and overstimulation, simplifying our lifestyles creates space for meaningful change. This isn't just about reducing material clutter; it's about giving our bodies and minds the chance to function as they were designed to. By letting go of habits, foods, and routines that disrupt our natural rhythms, we can tap into the wisdom of our biology, guided by both ancient traditions and modern science.

Fasting, for instance, is one of the most powerful ways to trigger autophagy. While modern society often views fasting as extreme, it has been a cornerstone of health practices for millennia. Across cultures and religions, fasting was revered not just for its spiritual benefits but also for its ability to cleanse the body and sharpen the mind. Science now confirms what ancient traditions understood intuitively: fasting activates autophagy, giving cells the space to repair, detoxify, and rejuvenate.

But autophagy isn't just about going without food. It's also about what you choose to put into your body. Modern diets, with their excess sugar, refined carbohydrates, and processed foods, often inhibit autophagy, overloading the system and accelerating aging and disease. This book explores how dietary shifts—toward nutrient-dense, whole foods—can support autophagy and help you achieve better health. Food is more than fuel; it's a message to your cells. Learning to send the right signals is key to optimising your body's natural repair systems.

The science of autophagy, while groundbreaking, is beautifully simple. Your body knows how to heal—it just needs the right conditions. Fasting, exercise, and even certain supplements can trigger this repair mechanism, allowing you to detoxify at a cellular level. Think of it as your body's spring cleaning: old, damaged materials are cleared away, making room for new growth and efficiency. This process doesn't just help with aging gracefully; it also reduces the risk of chronic diseases, from diabetes to neurodegenerative disorders.

Throughout this book, you'll find practical tools and strategies to activate and sustain autophagy. Central to this is the **NoMAD Plan**, a structured yet adaptable approach to simplifying health management. The plan integrates fasting protocols, dietary adjustments, and innovative tools like nanocelle nutrients to make autophagy accessible to everyone. It's not a one-size-fits-all solution but a flexible framework that meets you where you are and guides you toward where you want to be.

Beyond the science, the book brings this journey to life through real-world stories. These inspiring examples demonstrate how people from all walks of life have transformed their health by embracing simplicity and activating autophagy. From managing chronic conditions like diabetes and ADHD to achieving better energy and focus, these stories illustrate the profound potential of this approach.

While the focus is on the individual, the implications of autophagy go beyond personal health. By adopting these principles, we contribute to a broader culture of sustainability—one that respects the balance of nature and minimises unnecessary consumption. The simplicity of autophagy isn't just good for your cells; it's good for the planet.

The journey outlined in this book is deeply rooted in the idea that health and vitality are not about quick fixes or extreme measures. They are about understanding and working with the natural rhythms of your body. By aligning your lifestyle with these rhythms—through fasting, mindful eating, and thoughtful self-care—you unlock the incredible power within you to heal, grow, and thrive.

This book isn't just a guide; it's an invitation to rediscover the elegance of simplicity and the resilience of your own biology. As you read, you'll learn not only about the mechanisms of autophagy but also about the joy of living in harmony with your body's natural wisdom. You'll discover how small, intentional steps can lead to profound change, how simplicity can create space for growth, and how embracing the power of less can help you achieve more.

Let this be the beginning of your journey—a journey toward renewal, vitality, and a life that reflects the best version of yourself. In a world of noise, choose clarity. In a world of excess, choose simplicity. And in a world that often pulls you away from yourself, choose to come home to the incredible healing power within.

The Autophagy Path to Healing

Summary: Introduction

Chapter 1: The Power of Less: A Philosophy of Simplification

Imagine your life as a garden. Each plant in the garden has its own purpose—some grow food, others provide shade, and some simply add beauty. But if the garden becomes overcrowded, the plants compete for sunlight, water, and nutrients, and the entire ecosystem suffers. This is a metaphor for modern life: overburdened with commitments, possessions, and distractions, we lose sight of what truly matters.

The philosophy of simplification is not about deprivation but liberation. It is about removing the unnecessary to make space for what is meaningful, valuable, and fulfilling. Across cultures, history, and contemporary lives, the principle of "less is more" has been practiced by some of the world's most inspiring individuals and communities. Their stories demonstrate that simplicity fosters clarity, purpose, and well-being. This chapter explores how embracing simplicity shapes our mental, emotional, and physical lives, with lessons from cultural traditions, nature, and famous figures who embody this practice.

The Case for Less

Modern society glorifies abundance. Success is often measured by how much we own, achieve, or consume. Yet research and experience reveal the opposite: excess leads to stress, decision fatigue, and discontent—a phenomenon psychologists call the **paradox of choice**. The more we accumulate, the harder it becomes to focus on what truly matters, leaving us overwhelmed and unsatisfied.

The constant influx of options, from streaming services to career paths, taxes our cognitive resources. Studies show that excessive choices lead to dissatisfaction, even when outcomes are favorable. This highlights a central truth: life's richness isn't measured by how much we have but by the value we derive from it.

Intentional scarcity, on the other hand, fosters clarity. It forces us to prioritise and cherish the essentials. History is full of individuals and communities who embraced this principle, using simplicity as a tool to amplify their focus, resilience, and creativity.

Lessons from Cultural Simplicity

Hiiggi: Spiritual Minimalism

The Hiiggi, an ethnic group in Nigeria, exemplify spiritual minimalism through their philosophy of balance and simplicity. Their lives center on subsistence farming, communal living, and rituals that honour nature. For the Hiiggi, possessions are functional rather than excessive, and their world view prioritises harmony over accumulation. Their practices echo a universal truth: simplicity isn't about what we lack but what we gain by focusing on essentials.

Maasai: Intentional Living

The Maasai of East Africa are renowned for their simple yet resilient way of life. Their economy, based on livestock, prioritises sustainability and communal sharing. Maasai culture thrives on minimalism, with homes constructed from natural materials and possessions limited to what is necessary. This intentional lifestyle fosters a profound connection to nature and community, reminding us that abundance is not measured in possessions but in relationships and purpose.

Aboriginal Wisdom: Living with the Land

Australia's Aboriginal peoples, one of the world's oldest continuous cultures, embody a lifestyle of deep simplicity. Their "Dreamtime" philosophy connects them to the land, emphasising sustainable living and spiritual harmony. Aboriginal communities often rely on what the land naturally provides, teaching modern society the importance of living lightly and respecting ecological limits. Their way of life demonstrates that true abundance lies in alignment with nature, not in overexploitation.

Blue Zone Populations: Longevity Through Simplicity

The world's Blue Zones—regions where people live significantly longer than average—share a common thread: simplicity. Whether it's the plant-based diets of Okinawans in Japan, the active, community-driven lives of Sardinians in Italy, or the faith-centered routines of Seventh-Day Adventists in Loma Linda, California, these populations prioritise natural living, social connections, and purposeful work. Their longevity offers evidence that simplicity isn't just a philosophy; it's a key to health and happiness.

Scandinavian Hygge: The Art of Cozy Contentment

In Denmark, the cultural concept of **hygge** reflects simplicity as emotional warmth and contentment. Unlike the Swedish **lagom**, which centers on balance, hygge focuses on creating cozy environments and fostering connections. Whether through candlelit gatherings or quiet moments with a book, hygge transforms ordinary experiences into sources of joy. This practice highlights that happiness is often found in the smallest, simplest pleasures.

Famous Figures Who Embody Simplicity

Billionaires Who Choose Simplicity

The perception of billionaires often conjures images of lavish lifestyles, private jets, and sprawling mansions. Yet some of the wealthiest people in the world defy this stereotype, embracing simplicity and rejecting excess.

- **Warren Buffett:** Despite being worth around $150 billion, Buffett has lived in the same modest Nebraska home he purchased for $30,000 over 60 years ago.

Known for his unassuming lifestyle, Buffett exemplifies how simplicity fosters focus and clarity in both personal and professional life.

- **Chuck Feeney:** Once worth $7.5 billion, Feeney donated most of his wealth to charity, leaving himself with just $2 million—a deliberate choice to live simply and impactfully.
- **David Cheriton:** A Stanford professor and early Google investor, Cheriton avoids extravagance, lives in the same house he has owned for 30 years and even cuts his own hair.
- **Azim Premji:** One of India's wealthiest men, Premji is known for his frugality, from flying economy class to monitoring office expenses like toilet paper.
- **Carlos Slim Helú:** The Mexican telecom magnate worth $80 billion lives in a six-bedroom house he has owned for decades and drives himself to work.

The Benefits of Simplified Living

Embracing simplicity yields profound benefits:

- **Enhanced Clarity and Creativity:** Removing distractions amplifies focus, as seen in leaders like Steve Jobs and Albert Einstein.
- **Stronger Relationships:** Simplified living fosters connection over consumption, exemplified by the Amish and Scandinavian cultures.
- **Environmental Sustainability:** Living lightly, like migratory birds or honeybees, reduces our ecological footprint.
- **Deeper Fulfillment:** Practices like **hygge** and mindfulness cultivate gratitude and emotional well-being.

The Autophagy Path to Healing

Embracing the Power of Less

Simplicity isn't about sacrifice—it's about freedom. It allows us to focus on what truly matters, creating space for growth, connection, and fulfillment. By learning from nature, cultures, and individuals who embody this philosophy, we can see that living with less doesn't mean lacking—it means thriving.

As Gandhi, Jobs, and Thoreau have shown, the power of less clears away the noise, allowing us to focus on life's essence. Whether addressing personal clutter or systemic challenges like healthcare, the philosophy of simplification offers a path to clarity, sustainability, and peace.

Summary: The Power of Less: A Philosophy of Simplification

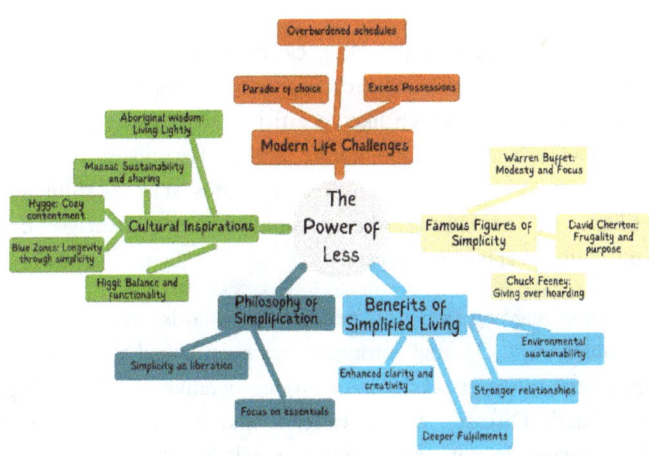

Chapter 2: Autophagy Unveiled

2.1 What is Autophagy?

Autophagy, derived from the Greek term meaning "self-eating," might sound unsettling at first, but it is one of the most fascinating and essential processes in our bodies. Think of autophagy as your body's internal recycling system, a housekeeping service for your cells that is operational 24/7.

Imagine your body as a bustling metropolis and each cell as a home within this vast city. Over time, every home accumulates clutter—broken furniture, worn-out appliances, and general waste. Without regular cleaning, the house becomes uninhabitable, and its functionality suffers. Similarly, our cells accumulate "cellular junk," including damaged organelles, misfolded proteins, and other debris that hinder their optimal functioning.

Autophagy acts as the city's cleaning crew, identifying and removing this waste while recycling useful components. It is a continuous process, much like an efficient municipal service that keeps the streets clean and homes livable. Beyond its role as a cellular janitor, autophagy ensures that cells stay youthful and resilient by preventing the accumulation of damage that contributes to aging and disease.

This ongoing renewal process safeguards our health in profound ways. Optimal autophagy has been linked to preventing age-related diseases such as neurodegenerative disorders (e.g., Alzheimer's Disease), certain cancers, and metabolic conditions like diabetes. It not only cleans our cellular "homes" but also fortifies the foundation, enabling us to maintain vitality and energy as we age.

The Autophagy Path to Healing

2.2 The History and Evolution of Autophagy

Autophagy is far from a modern discovery. It is an ancient survival mechanism, deeply embedded in the evolutionary history of life on Earth. Its origins date back millions of years, to a time when our ancestors faced unpredictable cycles of feast and famine.

To understand this better, imagine living in a harsh wilderness where the food supply is uncertain. During times of scarcity, you would become creative—perhaps repurposing leftovers, fixing broken tools, or stretching resources until the next meal. Autophagy mirrors this resourcefulness at a cellular level. In periods of nutrient scarcity, cells dismantle and recycle their non-essential components to fuel critical processes. This ensures survival even when external resources are scarce.

Beyond nutrient scarcity, autophagy also evolved as a response to other stresses, such as infections or physical injuries. For example, when an infection strikes, autophagy clears out damaged or infected cellular components, much like isolating and dismantling a malfunctioning machine in a factory to prevent a system-wide breakdown. This capacity for self-repair has been a cornerstone of human adaptability and resilience.

Over time, this remarkable process has been fine-tuned to support not only survival but also optimal functioning during periods of abundance. Whether dealing with oxidative stress, inflammation, or the wear and tear of aging, autophagy has remained a steadfast ally in our biological toolkit.

2.3 Mechanisms of Autophagy

To truly grasp autophagy, we must delve into the intricate world of our cells. Picture each cell as a self-contained factory. Over time, machinery within the factory can break down, tools may become dull, and waste begins to pile up. Autophagy is the factory's maintenance team, equipped with the ability to dismantle old equipment and repurpose its materials to build new, functional tools.

The process begins with the formation of an **autophagosome**, a double-membrane structure that wraps around cellular waste. This structure is akin to a garbage truck collecting refuse from the factory floor. Once the autophagosome is loaded, it merges with a lysosome, the cell's recycling plant. Inside the lysosome, powerful enzymes break down the waste into its basic components—proteins, lipids, and carbohydrates—which are then reused to build new cellular machinery.

This meticulous recycling process extends to the mitochondria, the cell's energy powerhouses. Mitochondria naturally produce reactive oxygen species (ROS), by-products that can damage cellular components if left unchecked. Autophagy cleans up damaged mitochondria and eliminates excess ROS, ensuring that these power stations run efficiently and generate clean energy.

An additional benefit of autophagy lies in its impact on insulin regulation. In Western diets, where insulin resistance is increasingly common, fasting-induced autophagy plays a critical role in restoring balance. Think of insulin as a key that unlocks the door to cellular energy storage. In insulin resistance, the lock becomes faulty, requiring more keys (insulin) to open it. Autophagy repairs this mechanism by improving insulin sensitivity, reducing fasting insulin levels, and curbing the cycle of fat storage, sugar cravings, and inflammation.

During fasting, autophagy also triggers the release of growth hormone, promoting fat breakdown and toxin elimination. This detoxification process is particularly vital, as many environmental toxins accumulate in fat tissue. By clearing out these toxins, autophagy not only enhances metabolic health but also supports overall cellular rejuvenation.

2.4 Autophagy and Disease

When autophagy functions optimally, it acts as a shield against numerous health challenges. However, when this process is impaired, the consequences can be far-reaching.

A breakdown in autophagy has been implicated in the development of various diseases, including:

- **Heart disease:** Accumulation of damaged mitochondria in cardiac cells contributes to poor heart function.
- **Diabetes and fatty liver:** Impaired autophagy disrupts insulin signaling and fat metabolism.
- **Neurodegenerative diseases:** The buildup of toxic proteins, such as tau in Alzheimer's, is linked to deficient autophagic activity.
- **Cancer:** Dysfunctional autophagy allows damaged DNA and abnormal cells to persist, increasing cancer risk.
- **Accelerated aging:** The inability to clear cellular debris accelerates wear and tear on tissues and organs.

In essence, autophagy serves as the body's quality control system. When it falters, the cellular "homes" become cluttered and dysfunctional, paving the way for chronic illnesses and premature aging.

2.5 Balancing Autophagy and Growth

Autophagy operates in a delicate balance with **mTOR** (mammalian target of rapamycin), a key regulator of cell growth and reproduction. While autophagy represents the catabolic (breakdown and repair) phase, mTOR drives the anabolic (growth and building) phase.

Think of mTOR as the foreman of a construction site. Its job is to build new structures and expand the city. However, continuous construction without periodic cleanup leads to overcrowding and chaos. Similarly, excessive mTOR activation—

triggered by overeating, particularly protein-rich diets—can result in harmful protein accumulation, as seen in Alzheimer's disease.

Fasting and caloric restriction help tip the balance toward autophagy, allowing the city to reset and repair. The activation of **sirtuins**, a family of proteins linked to longevity, further amplifies autophagic activity. Sirtuins act as guardians of cellular health, repairing DNA, enhancing stress resistance, and promoting metabolic efficiency.

One such enzyme, AMP kinase, plays a pivotal role in regulating energy balance. Found abundantly in the liver, brain, and skeletal muscles, AMP kinase is activated by both exercise and fasting, making it a key player in the symphony of autophagy.

2.6 The Bigger Picture

Autophagy is more than just cellular cleanup—it is the foundation of our body's resilience and longevity. By maintaining a steady cycle of breakdown and renewal, it ensures that our cells operate at peak efficiency.

Imagine a well-tuned orchestra where every instrument plays in harmony. Autophagy acts as the conductor, orchestrating a balance between growth and repair, ensuring that the body's symphony remains melodious. When the process falters, the music becomes chaotic, leading to dysfunction and disease.

Understanding autophagy empowers us to make lifestyle choices that support this critical process. Intermittent fasting, caloric restriction, and regular exercise are some of the most effective ways to activate autophagy and optimise our health.

2.7 Autophagy: The Key to Thriving

By unlocking the secrets of autophagy, we gain insight into one of nature's most ingenious mechanisms for self-preservation. From the evolutionary lessons of our ancestors to the cutting-edge science of today, autophagy serves as a reminder of the body's remarkable capacity for renewal.

As we embrace practices that enhance autophagic activity, we take control of our health and longevity. Whether through fasting, mindful eating, or physical activity, we become architects of our own biological destiny, crafting a life of vitality and resilience.

The **6 R Plan** of Levitas emphasises the critical role of detoxification in achieving optimal health. From the moment of conception, our bodies are exposed to a range of environmental toxins, including heavy metals, pesticides, plastics, and endocrine disruptors, which can accumulate over time. These contaminants affect not just our own health but may even be inherited through maternal exposure during pregnancy. This phenomenon, often referred to as **transgenerational toxicity**, underscores the importance of proactive detoxification strategies to mitigate these risks.

Autophagy is a cornerstone of this detoxification process, enabling the body to clear out damaged proteins, dysfunctional organelles, and intracellular toxins. However, effective detoxification requires more than autophagy alone. The **6 R Plan** provides a structured approach, encompassing **Restore, Release, Repair, Renew, Reframe, and Represent**, to ensure comprehensive support for cellular and systemic detoxification.

- **Restore** focuses on building a strong foundation by replenishing essential nutrients, minerals, and hormones to optimise the body's infrastructure.
- **Release** addresses the elimination of toxins stored in fat cells and other tissues, often triggered during fasting or calorie restriction, and ensures their safe excretion.
- **Repair** emphasises the regeneration of damaged tissues and cellular structures, using the raw materials recycled through autophagy.

- **Renew** encourages the restoration of vitality through cellular rejuvenation, fostering a balance between growth and repair.
- **Reframe** involves adopting new mindsets and behaviors to minimise future toxic exposures, including dietary choices, stress management, and environmental awareness.
- **Represent** highlights the role of self-advocacy and community education in fostering healthier environments and promoting shared accountability for reducing collective toxic burdens.

By aligning autophagy-activating practices with these six principles, we create a holistic framework to detoxify and fortify the body against the challenges of modern living. Understanding that we are exposed to contaminants even before birth encourages us to prioritise detoxification at every stage of life. In doing so, we not only reclaim our health but also ensure that future generations inherit a cleaner, healthier world.

Summary: *Autophagy Unveiled*

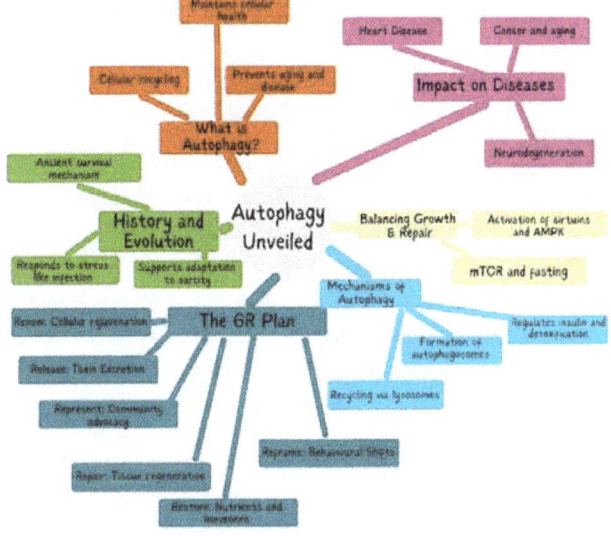

Chapter 3: Fasting: A Historical Journey Across Rituals, Cultures, and Science

Fasting has shaped human history as a multifaceted practice, blending spiritual, cultural, political, and medical dimensions. By delving into its evolution—from ancient philosophers like Hippocrates and Aristotle to the warrior tribes of Spartacus, from religious devotion to political defiance—we uncover how fasting has been a powerful tool for survival, discipline, protest, and healing. Modern science adds a new dimension to this narrative, particularly the concept of **hormesis**—how mild stressors, like fasting, can strengthen the body and mind early on but may become problematic if pushed too far.

Fasting in the Ancient World: The Philosophers and Tribes

Hippocrates: The Father of Medicine

Hippocrates, regarded as the father of medicine, was one of the earliest advocates of fasting for health. He famously stated, **"To eat when you are sick is to feed your illness,"** emphasising the body's natural capacity for self-healing. He believed fasting allowed the digestive system to rest and gave the body the energy needed to combat illness. This foundational idea persists in modern medicine, where fasting states like ketosis are explored for their therapeutic benefits.

Hippocrates also recommended fasting for conditions ranging from digestive disorders to epilepsy. He observed that many patients improved when they abstained from food, attributing their recovery to the body's innate wisdom rather than medicinal

The Autophagy Path to Healing

interventions. These observations laid the groundwork for viewing fasting as an active healing tool.

Aristotle and Philosophical Discipline

Aristotle, one of the greatest thinkers of ancient Greece, regarded fasting as a pathway to self-mastery. For him, controlling physical desires like hunger was essential for the mind's liberation. This discipline allowed individuals to transcend the animalistic nature of physical cravings and attain higher reasoning. Aristotle's thoughts influenced the development of Stoic philosophy, which emphasised enduring discomfort for greater intellectual and spiritual clarity.

Spartacus and the Warrior Tribes

In the ancient world, fasting was not only a philosophical or medical practice but also a survival strategy. The **Spartacus-led rebellion** (73–71 BCE) against Roman tyranny showcased the resilience of warriors who endured hunger as they fought for their freedom. These gladiators, often trained in Spartan-like austerity, incorporated fasting as part of their mental and physical conditioning.

The **Spartans**, legendary for their harsh discipline and militaristic culture, saw fasting as a tool to sharpen their senses and strengthen their resolve. Hunger was viewed not as a weakness but as an opportunity to test the body's limits. This belief echoed in the training of other warrior tribes, such as the Viking berserkers and the samurai of feudal Japan, who used controlled deprivation to heighten focus and prepare for battle.

Hormesis: The Sweet Spot Between Benefit and Risk

Fasting exemplifies **hormesis**, a biological principle where mild stressors stimulate beneficial adaptations. In the early stages of fasting, the body activates protective mechanisms, such as improved insulin sensitivity, enhanced mitochondrial function, and **autophagy**—the cellular recycling process that removes damaged components. These adaptations enhance resilience, promoting longevity and disease resistance.

However, hormesis relies on balance. Short-term fasting delivers benefits, but prolonged fasting can push the body beyond its capacity to recover. Once energy reserves are depleted, fasting transitions from a hormetic stressor to a harmful state, with muscle breakdown, weakened immunity, and organ failure as potential outcomes.

Reassurance About Autophagy

Autophagy, a hallmark of fasting, is a natural and beneficial process. It allows cells to recycle damaged proteins and organelles, clearing out toxins and promoting cellular renewal. This process peaks during fasting periods of 16-48 hours, providing a safe and effective window for health optimisation. Concerns about fasting-related harm often stem from extreme or prolonged fasting, not from intermittent or short-term fasting where autophagy thrives.

For most healthy individuals, engaging in fasting protocols such as can safely activate autophagy without pushing the body into a negative stress zone. The approaches discussed are always with medical supervision and personalised

The Science of Survival Without Food and Water

How Long Can the Human Body Survive Without Food?

The length of time a person can survive without food depends on various factors, including body fat stores, muscle mass, hydration, and overall health. Generally, most individuals can survive **30 to 60 days** without food, provided they have access to water. Historical examples, such as the **Irish hunger strikes**, provide real-world data. Bobby Sands, an Irish activist, survived **66 days** without food before succumbing to organ failure. His case reflects the body's incredible ability to adapt to prolonged fasting, though the eventual consequences of starvation are irreversible.

The Autophagy Path to Healing

The Transition Stages During Prolonged Fasting

When the body is deprived of food, it undergoes distinct metabolic transitions to survive:

1. **Glycogen Depletion**

    a .The body first uses glycogen stored in the liver and muscles for energy. Glycogen stores last a day or so depending **on activity levels.**

    b. Blood sugar levels are maintained initially, but as glycogen reserves deplete, the body begins transitioning to alternative energy sources.

2. **Ketosis (day 2 onwards):**
    a. After glycogen is exhausted, the body turns to fat stores, breaking them down into **ketones** for energy. This process is called **ketosis.**
    b. Ketones fuel the brain, muscles, and vital organs, reducing the body's reliance on glucose.

3. **Autophagy Activation (24- 72 hours):**

    a. During this stage, cells begin breaking down old proteins and damaged

    b. organelles. This self-cleaning mechanism supports health and longevity.

4. **Starvation Mode (day 7 plus without support)**
    a. As fat stores are depleted, the body begins breaking down muscle protein to produce glucose through **gluconeogenesis**.
    b. At this stage, the body prioritises energy for critical functions, including the brain and heart, while conserving resources for survival.

5. **Terminal Starvation (after 30-60 days):**
   a. Once both fat and muscle reserves are critically low, organ failure occurs due to insufficient energy to sustain vital functions.
   b. Death is typically caused by cardiac arrest or infection due to immune suppression.

How Long Can the Human Body Survive Without Water?

Without water, survival is significantly shorter. On average, humans can survive only **3 to 7 days** without hydration, depending on environme   ntal factors like heat, humidity, and activity levels. Dehydration rapidly leads to electrolyte imbalances, kidney failure, and ultimately death.

Fasting as Therapy: Balancing Benefit and Caution

Modern clinics offering therapeutic fasting emphasise structured approaches to avoid the risks of over-fasting. For example, programs like those at the **Buchinger Wilhelmi Clinic** in Germany combine medical supervision with fasting protocols designed to activate autophagy and other health benefits while preventing harm.

It is important to note that fasting is not a one-size-fits-all solution. Individuals with chronic conditions, malnourishment, or a history of eating disorders should consult healthcare professionals before attempting fasting protocols. When practiced safely, fasting can harness the body's natural resilience and healing potential, exemplifying the delicate balance of hormesis.

A Timeless Practice for Body and Spirit

Fasting's universality reflects its profound impact on human history. From the **philosophical musings of Aristotle** to the rebellious spirit of **Spartacus**, from the spiritual quests of Native Americans to the scientific breakthroughs of autophagy, fasting has transcended cultures and eras.

The Autophagy Path to Healing

By balancing the principles of hormesis—leveraging the stress of fasting for renewal without overextending its demands—fasting remains a safe, transformative practice. Modern science reassures us that when done responsibly, fasting unlocks ancient wisdom while protecting health, empowering individuals to thrive both physically and spiritually.

Summary: Fasting: A Historical Journey Across Rituals, Cultures, and Science

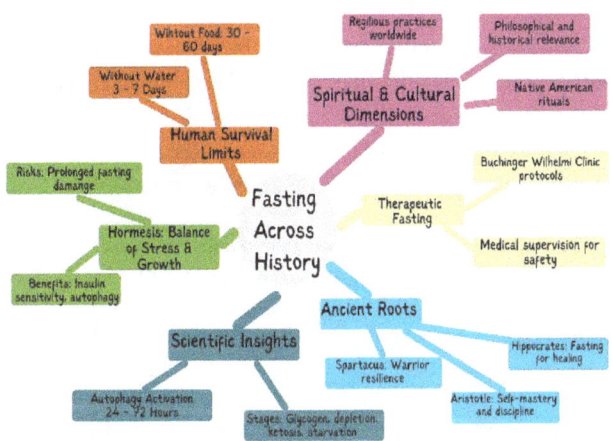

Chapter 4: Modern Diets and Autophagy

4.1 Overview of Popular Diets

In today's world, the sheer variety of diets reflects the growing awareness of how food impacts health, longevity, and performance. Diets are not just sources of sustenance; they influence complex biological processes, metabolic health, and overall well-being. One such critical process is autophagy, the body's natural mechanism for cleaning out damaged cells and regenerating healthier ones.

Autophagy, meaning "self-eating," is like a cellular recycling program that enables the body to maintain optimal function. The effectiveness of this process depends significantly on what, how, and when we eat. Different diets impact autophagy in unique ways, each offering insights into how food choices shape cellular health.

Let's explore a broader range of dietary approaches and how they influence autophagy and overall cellular health:

Atkins Diet

Focusing on low carbohydrate intake, the Atkins diet pushes the body to rely on fat for energy, inducing a state of ketosis. This metabolic switch encourages autophagy by mimicking the energy-scarce conditions typically seen during fasting. It is particularly effective in stimulating fat metabolism, cellular repair, and reducing insulin resistance.

Mediterranean Diet

Rich in whole foods, healthy fats, and antioxidants, the Mediterranean diet supports cellular health by reducing inflammation and oxidative stress. The emphasis on polyphenol-rich foods like olive oil, fish, nuts, and leafy greens creates an environment conducive to autophagy and longevity.

The Autophagy Path to Healing

Vegan Diet

A plant-based approach, the Vegan diet focuses on whole, nutrient-dense foods while avoiding animal products. By reducing intake of saturated fats and processed foods, this diet minimises cellular stress and promotes repair. Its high fiber content supports gut health, indirectly enhancing autophagy by maintaining a balanced microbiome.

Paleo Diet

Based on the presumed eating habits of our hunter-gatherer ancestors, the Paleo diet eliminates processed foods, grains, and dairy, emphasising whole foods like meat, fish, vegetables, and nuts. This approach helps reduce inflammation, improve insulin sensitivity, and support autophagy by minimising modern dietary stressors.

Ketogenic Diet

Being like the Atkins Diet but stricter in its macronutrient ratios, the Ketogenic diet forces the body into ketosis by drastically limiting carbohydrates. This diet enhances autophagy through energy scarcity, triggering the body to clear damaged cells and optimise metabolic pathways.

Intermittent Fasting (IF)

More of an eating pattern than a diet, intermittent fasting alternates between periods of eating and fasting. By mimicking nutrient deprivation, IF is one of the most effective ways to trigger autophagy. It supports metabolic flexibility, reduces inflammation, and promotes cellular repair across the body.

Blood Type Diet

This diet tailors food choices based on an individual's blood type, claiming that specific foods are more compatible with certain blood types. While scientific evidence supporting this approach is limited, its emphasis on whole foods and elimination of processed options can indirectly benefit autophagy.

Warrior Diet

The Warrior Diet involves eating small amounts of raw fruits and vegetables during the day and consuming one large meal at night. This pattern mimics periods of fasting followed by feasting, potentially triggering autophagy during the fasting window while supplying nutrients for repair in the eating phase.

Carnivore Diet

Focused solely on animal products, the Carnivore diet eliminates all plant-based foods. While controversial, this diet reduces carbohydrate intake to almost zero, potentially inducing ketosis and stimulating autophagy. However, the long-term impact of such an extreme dietary restriction requires further study.

Whole30 Diet

A 30-day elimination diet that avoids processed foods, added sugars, grains, legumes, and dairy, Whole30 emphasises clean eating. By reducing dietary stressors, this approach supports gut health and indirectly promotes cellular repair and autophagy.

Dash Diet

Designed to lower blood pressure, the Dash diet emphasises fruits, vegetables, whole grains, and low-fat dairy while limiting sodium and unhealthy fats. Its nutrient-dense focus supports cellular health, though its impact on autophagy is more indirect compared to ketogenic or fasting-based diets.

Raw Food Diet

Focusing on uncooked and minimally processed foods, the Raw Food diet emphasises high nutrient density and a low-calorie intake. While the raw aspect may not directly influence autophagy, the reduction in processed foods and emphasis on natural ingredients supports cellular health.

Low-FODMAP Diet

Primarily aimed at managing irritable bowel syndrome (IBS) and other digestive issues, the Low-FODMAP diet reduces fermentable carbohydrates. While its primary focus is gut health, reducing digestive stress may indirectly promote autophagy by supporting a balanced microbiome.

Anti-Inflammatory Diet

Designed to reduce chronic inflammation, this diet incorporates foods like fatty fish, berries, turmeric, and leafy greens. By lowering systemic inflammation, it supports the conditions under which autophagy can thrive.

Detox Diets

Often incorporating periods of fasting, juice cleanses, or elimination of processed foods, Detox diets aim to reduce the body's toxic load. These diets may support autophagy by promoting energy scarcity and reducing inflammatory dietary inputs.

4.2 The Role of Diet in Supporting Autophagy

While each diet offers unique pathways to health, their ability to support autophagy often boils down to common principles: nutrient balance, energy scarcity, and reduction of cellular stressors. Low-carb and ketogenic diets, for example, mimic fasting states that trigger autophagy, while nutrient-dense diets provide the building blocks for repair.

- **Fasting and Meal Timing**: Regardless of diet, fasting windows activate the migrating motor complex (MMC) and trigger autophagy. This combination enhances gut health, reduces toxic burden, and clears out damaged cells.

- **Nutrient Quality**: Diets rich in antioxidants and healthy fats, like the Mediterranean and Anti-Inflammatory diets, protect cells from oxidative damage, allowing autophagy to function more effectively.
- **Personalisation**: No single diet suits everyone. Factors like genetics, lifestyle, and health goals influence how well a diet supports autophagy, making personalised approaches critical.

Metabolic Plasticity and the Need for Variation

Metabolic plasticity—the body's ability to adapt its metabolism based on environmental and dietary inputs—is a cornerstone of health and survival. In nature, food availability fluctuated with seasons, migrations, and environmental changes, compelling our ancestors to adapt their diets and metabolic strategies. This inherent variability promoted resilience by triggering periods of energy abundance followed by scarcity, which activated processes like autophagy. Modern diets often lack this variation, leading to metabolic stagnation, overnutrition, and chronic diseases. To mimic natural patterns, incorporating dietary variation—such as alternating between higher-carb and low-carb days or including fasting periods—can restore metabolic flexibility. This approach aligns with the body's evolutionary design, fostering optimal cellular function, energy efficiency, and longevity.

This dynamic adaptation highlights the importance of not only what we eat but how we alternate and time our dietary practices to better align with the body's natural rhythms and capacity for self-repair.

4.3 Fasting: The Universal Autophagy Trigger

Across all dietary approaches, fasting remains the most potent activator of autophagy. It creates energy scarcity, forcing the body to prioritise essential functions and repair. Even diets with occasional indulgences can benefit from incorporating fasting to trigger cellular cleanup and enhance metabolic health.

The Autophagy Path to Healing

4.4 The Levitas Perspective

At Levitas Clinic, we recognise the individuality of health journeys. Our **NoMAD Plan** blends insights from diverse dietary approaches, emphasising structured fasting and nutrient-dense meals to maximise autophagy.

Imagine your body as a finely tuned orchestra. Each diet plays a unique instrument, contributing to the harmony of cellular health. The NoMAD Plan acts as the conductor, integrating these elements into a personalised, balanced approach that empowers your body to thrive.

Conclusion

Modern diets offer varied paths to health, but their impact on autophagy depends on their ability to balance nutrient density with periods of energy scarcity. Regardless of the diet, integrating fasting and focusing on whole, unprocessed foods creates the ideal conditions for autophagy to flourish. By understanding and tailoring your approach, you can harness the power of autophagy to achieve long-term health and vitality.

Summary: Modern Diets and Autophagy

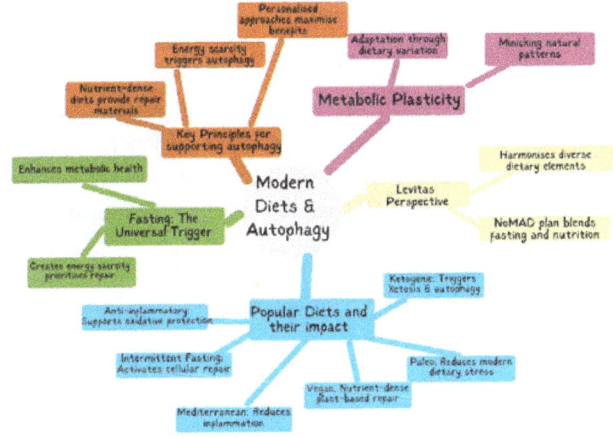

Chapter 5: The Science Behind Autophagy

Autophagy is an essential biological process that helps maintain cellular health by clearing away damaged components and regenerating new ones. It is like a built-in cleaning and recycling system that keeps our bodies functioning smoothly and efficiently. By understanding how autophagy works and how we can support it, we unlock powerful tools to improve our health, slow aging, and even prevent disease.

5.1 Cellular Restoration and Autophagy

Imagine your body as a busy city, where each cell is like a house. Over time, these houses accumulate broken furniture, faulty appliances, and clutter. Autophagy is the professional cleaning crew that clears out the waste, fixes what can be repaired, and ensures that everything functions as it should.

This process of cellular cleanup and renewal is essential for keeping your body running at its best. By removing old and damaged parts, autophagy prevents the buildup of toxic materials that can lead to aging and diseases like cancer, Alzheimer's, and heart disease. It also recycles materials, using them to create new, healthy components that keep cells strong and resilient.

Without autophagy, the buildup of cellular waste can overwhelm the body, leading to a loss of efficiency, energy, and function. But when autophagy works optimally, it is like a constant home renovation, keeping everything fresh and functional.

Key Takeaway: Autophagy helps your cells stay youthful and healthy, reducing the risk of disease and extending the time you spend in good health.

The Autophagy Path to Healing

5.2 The Trigger Mechanisms

Autophagy doesn't happen constantly—it is activated under specific conditions. The most powerful triggers are fasting, exercise, and nutrient changes. These mild stresses encourage the body to clean up and repair itself, preparing for future challenges.

- **Fasting**: When you stop eating for a while, your body shifts into survival mode, using stored energy and breaking down old, damaged parts of cells to recycle resources. This helps keep cells efficient and reduces waste.
- **Exercise**: Physical activity causes small amounts of stress and minor damage to muscle cells, which signals the body to repair and strengthen them. This process also activates autophagy, improving overall cellular health.
- **Nutrient Signals**: Reducing protein or carbohydrate intake can also activate autophagy by mimicking the effects of fasting. For example, low-carb diets or calorie restrictions stimulate repair mechanisms in the body.

Analogy: Think of a factory running out of new materials to work with. Instead of shutting down, it reuses and repurposes old, unused parts to keep running efficiently. This is how autophagy keeps your body functioning even under stress.

Key Takeaway: Fasting, exercise, and diet changes are simple ways to trigger autophagy and support your body's natural repair systems.

5.3 The Work of Dr. Yoshinori Ohsumi

Dr. Yoshinori Ohsumi won the Nobel Prize in 2016 for discovering the mechanisms of autophagy. His research revealed how this process works at a cellular level and showed its importance in preventing disease and slowing aging.

The Autophagy Path to Healing

Dr. Ohsumi found that autophagy is like the janitors in a building—not only keeping it clean but preventing it from falling apart. When autophagy is working well, it protects cells from damage, clears harmful proteins, and ensures proper function. His discoveries opened the door to new ways of thinking about health and aging, showing how we can use autophagy to stay healthier for longer.

Key Takeaway: Dr. Ohsumi's research highlights how critical autophagy is to health, especially in today's world of constant food availability and limited physical activity.

5.4 Autophagy's Role in Disease Prevention

Autophagy plays a vital role in protecting against many diseases:

- **Cancer**: By clearing damaged parts of cells, autophagy helps prevent the mutations that can lead to cancer.
- **Neurodegenerative Diseases**: Conditions like Alzheimer's and Parkinson's occur when harmful proteins build up in the brain. Autophagy clears these proteins, protecting brain health.
- **Heart Disease**: Autophagy supports healthy heart cells by reducing inflammation and removing damaged mitochondria.
- **Diabetes and Obesity**: By regulating how the body uses fats and sugars, autophagy helps maintain a healthy metabolism.

Key Takeaway: Supporting autophagy can reduce the risk of many major diseases by keeping cells clean and functional.

5.5 Autophagy and Anti-Aging

Aging is caused by the gradual breakdown of cells and tissues, leading to loss of function, disease, and frailty. Autophagy slows this process by keeping cells clean, efficient, and adaptable. It addresses many of the biological factors behind aging, known as the **12 Hallmarks of Aging**:

1. **Genomic Instability**: Autophagy reduces damage to DNA by removing sources of oxidative stress.
2. **Telomere Attrition**: While it doesn't stop telomere shortening, autophagy protects cells from additional stress that speeds up the process.
3. **Epigenetic Changes**: Autophagy helps regulate gene expression by maintaining a healthy cellular environment.
4. **Loss of Proteostasis**: It clears harmful protein aggregates that accumulate with age.
5. **Deregulated Nutrient Sensing**: Fasting and nutrient changes reset metabolic pathways, improving autophagy.
6. **Mitochondrial Dysfunction**: Autophagy replaces damaged mitochondria with healthier ones, improving energy production.
7. **Cellular Senescence**: It helps remove aging cells that no longer function properly.
8. **Stem Cell Exhaustion**: Autophagy protects stem cells from damage, preserving their regenerative potential.
9. **Altered Communication**: It reduces inflammation, restoring balance in cellular signaling.
10. **Chronic Inflammation**: By clearing damaged parts of cells, autophagy reduces the sources of inflammation.
11. **Impaired Autophagy**: As we age, autophagy slows down. Supporting it through fasting and exercise can counteract this.
12. **Microbiome Changes**: Autophagy helps maintain gut health by supporting healthy cells in the intestinal lining.

Key Takeaway: Autophagy addresses many causes of aging, making it a key process for staying youthful and healthy.

5.6 Supporting Autophagy in Daily Life

The good news is that we can support autophagy with simple lifestyle changes:

- **Fasting:** Even short fasting periods (12-16 hours) can activate autophagy and support cellular repair.
- **Exercise:** Regular activity, particularly high-intensity or endurance training, triggers autophagy in muscle cells.
- **Diet Choices:** Low-carb or calorie-restricted diets mimic fasting conditions and boost autophagy.
- **Sleep:** Quality sleep allows the body to repair itself, supporting autophagy and overall health.
- **Supplements:** Certain compounds like resveratrol, spermidine, and green tea extract may enhance autophagy.

Conclusion

Autophagy is a natural process that holds the key to better health, disease prevention, and slowing aging. By understanding how it works and supporting it through fasting, exercise, and healthy living, we can unlock its full potential. In doing so, we give our bodies the tools they need to clean, repair, and thrive for years to come.

The science behind autophagy is a reminder that sometimes, less is more. By embracing simplicity—whether it's eating less often, moving more, or prioritising rest—we can activate our body's incredible ability to renew itself, ensuring a longer and healthier life.

The Autophagy Path to Healing

Summary: The Science Behind Autophagy

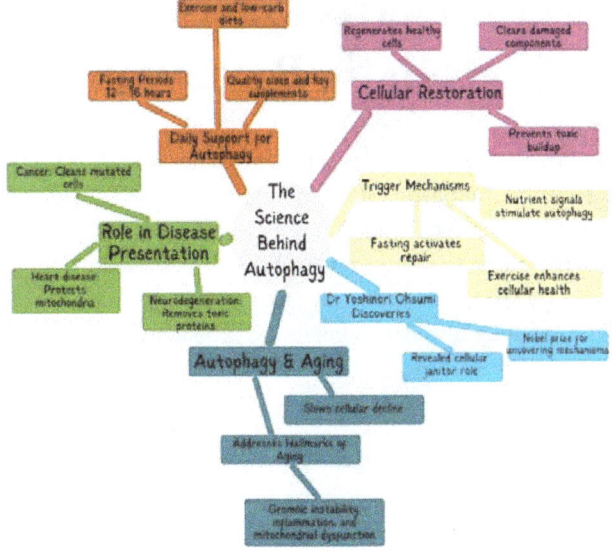

Chapter 6
The NoMAD Autophagy Plan: Detoxification and Holistic Renewal

Now that we understand the concept of autophagy, let us explore the NoMAD Autophagy Plans. These personalised programs are carefully built on the foundation of fasting and medical supervision to ensure safe and effective results. Designed to adapt to individual needs, the plans leverage advanced technologies and flexible delivery options, including home-based programs, wellness centers, and immersive retreats.

The NoMAD Autophagy Plan combines fasting, plant-based nutrition, advanced supplementation, and innovative delivery technologies such as nanocelle solutions and nutrient-rich drips. This approach activates and supports key detoxification pathways—including the gut, liver, kidneys, lymphatic system, and skin—to promote optimal health and cellular repair. The plans are designed to be flexible and customisable, ensuring they meet the unique needs of each participant while utiliSing cutting-edge health technologies.

NoMAD's Phases for Holistic Detoxification

The plan revolves around two synergistic phases:

1. 3-Day Water Fast

This phase is designed to initiate autophagy and stimulate deep cellular cleanup. It focuses on recycling damaged proteins, enhancing mitochondrial efficiency, and reducing inflammation. Mineral support through hypertonic solutions and trace minerals like Celtic salts is critical during this phase.

2. 10-Day Plant-Based Diet

Following the water fast, this phase sustains detoxification while replenishing the body with vital nutrients. The emphasis is on nutrient-dense, mineral-rich foods to restore and fortify the body's reserves.

The Challenges

Why Illness Can Arise During Fasting

While fasting is an ancient practice with profound benefits, it is important to recognise that the body can experience challenges during this process, particularly related to **mineral loss**. As the body transitions into a fasting state, it begins to use stored energy and initiates deep detoxification. This increased metabolic activity can lead to the depletion of essential minerals, which are crucial for maintaining hydration, nerve function, and cellular processes.

Symptoms such as fatigue, dizziness, muscle cramps, or even nausea can occur when minerals like sodium, potassium, magnesium, and calcium are lost through sweat, urine, or increased metabolic demand. This phenomenon underscores the importance of **supporting the body's mineral balance** during fasting to ensure the process is safe and effective.

How We Mitigate Mineral Loss During a Water Fast

To address the risk of mineral depletion and prevent illness during fasting, the NoMAD Autophagy Plan includes targeted strategies compatible with water fasting:

1. **Celtic Salts**
    a. A small amount of Celtic salt, dissolved in water, is used to provide trace minerals and replenish electrolytes.
    b. This method ensures hydration and mineral balance without interrupting the fasting state.

2. **Hypertonic Solutions**
   a. Marine plasma-based hypertonic solutions are incorporated to deliver a concentrated source of bioavailable minerals.
   b. These solutions are taken in minimal quantities to avoid breaking the fast while supporting cellular and organ function.
3. **IV Drips**
   a. For participants who may experience heightened mineral loss or other challenges during fasting, IV drips are available as part of the NoMAD plan.
   b. These drips deliver essential minerals and hydration directly into the bloodstream, ensuring efficient replenishment without interfering with the fasting process.
   c. This added layer of support is one of the unique benefits of the NoMAD Autophagy Plan, enabling participants to sustain the benefits of fasting while mitigating potential risks.

Other Challenges

Gender-Specific Fasting Variations and Optimal Windows

The NoMAD Autophagy Plan also recognises that fasting is not a one-size-fits-all process, and the timing and duration of fasting are carefully tailored to suit the unique biological needs of men and women.

- **For Women:**
  Women's hormonal cycles play a critical role in determining how the body responds to fasting. During certain phases of the menstrual cycle, fasting may be less effective or even counterproductive due to hormonal fluctuations that influence energy demands and metabolic stress. The NoMAD Plan adjusts fasting windows to align with times when the body is better equipped to handle the metabolic shifts of fasting, ensuring optimal

outcomes while minimizing disruptions to hormonal balance.

- **For Men**:
  Men generally have more consistent hormonal patterns, allowing for greater flexibility in fasting schedules. However, fasting windows are still customized based on activity levels, energy demands, and individual health goals to maximise the benefits of autophagy and cellular repair.

Measuring Success:

Glucose-Ketone Index and Ketosis Levels

To objectively measure the success of fasting and autophagy activation, the NoMAD Plan uses the **Glucose-Ketone Index (GKI)**. This index is calculated by dividing blood glucose levels by blood ketone levels, providing a simple and reliable indicator of metabolic state.

- **Optimal Ketosis for Autophagy**:
  A GKI of **1.0 or lower** suggests the body has entered a state of deep ketosis, where autophagy is likely optimised. This is the target range for most participants during the fasting phase, as it indicates that the body is effectively burning fat for fuel and engaging in cellular repair.
- **Monitoring and Adjustments**:
  Regular measurements of glucose and ketones allow us to monitor progress and make adjustments as needed. For participants struggling to achieve the desired GKI, interventions such as IV drips or adjusted fasting windows may be introduced to support the process.

1. **Glycogen Depletion**

   Initially, the body depletes glycogen stores in the liver and muscles. Glycogen holds water, so its breakdown results in a noticeable drop in weight, mostly due to water loss. This phase is essential for autophagy to occur, as the removal of glycogen signals the body to shift into a fasting state.

2. **Protein Debris Clearance**
After glycogen is depleted, the body begins recycling damaged proteins and cellular debris. This is the key mechanism of autophagy, where the body removes dysfunctional components to make way for regeneration.

3. **Fat Biohacking**
Once the body transitions fully into ketosis, fat becomes the primary fuel source. This phase is marked by a more stable weight loss, as stored fat is metabolised and used for energy. This is also when autophagy reaches its peak efficiency, as the body operates in a clean, energy-efficient state.

What is the The Added Benefit of NoMAD

The NoMAD Autophagy Plan stands apart because it integrates cutting-edge technologies like IV drips and nanocelle solutions to address the unique challenges of fasting while enhancing its benefits. By combining BCI monitoring, GKI tracking, gender-specific fasting schedules, and advanced interventions, the NoMAD Plan ensures participants achieve deep autophagy and long-lasting health renewal

How NoMAD Helps You Achieve Success

- **Tailored Timing**: The NoMAD Plan adjusts fasting schedules based on your body type and gender to ensure you reach the glycogen depletion stage as quickly and safely as possible.
- **Support Tools**: If you struggle to get into ketosis or face challenges like fatigue or mineral loss, tools like **IV drips** or **Celtic salts** help your body stay balanced.
- **Measuring Progress**: Using GKI and BCI ensures that your fasting is effective, allowing us to monitor your success and make adjustments if needed.

The Autophagy Path to Healing

How Drips and Nanocelle Technology Support the NoMAD Autophagy Plan

The NoMAD Autophagy Plan not only focuses on fasting to activate autophagy but also integrates advanced tools to maintain this state effectively and support recovery. Two essential elements of the plan are **nutrient drips** and **nanocelle technology**. These tools are designed to "feed the cell, not the stomach," ensuring optimal cellular function without disrupting the fasting state.

1. Drips to Maintain Autophagy

During fasting, it's crucial to maintain autophagy—the body's natural process of recycling damaged proteins and cellular debris—while avoiding anything that might disrupt this delicate state.

- **How Drips Help Maintain Autophagy**:
    o Certain drips provide **minerals** (e.g., magnesium, potassium) and hydration without introducing calories that could interrupt autophagy.
    o **Electrolyte Support**: Balances essential minerals lost during fasting, reducing symptoms like fatigue or cramps.
    o **Immune Nutrients**: Ingredients like zinc, selenium, and vitamin C can be included in small, fast-compatible amounts to support immune function during fasting without disrupting cellular cleanup.

By ensuring that the body's mineral and hydration needs are met, these drips allow participants to stay in the autophagic state longer, enhancing the detoxification and repair process.

2. Drips to Support Rebuilding After Fasting

Once the fasting phase is complete and autophagy has done its work, the focus shifts to rebuilding and replenishing the body. This phase is essential for restoring energy, strengthening immunity, and promoting regeneration.

The Autophagy Path to Healing

- **How Drips Help with Recovery**:
    - **Amino Acids**: Key building blocks for protein synthesis, supporting muscle repair, and cellular regeneration.
    - **Immune Nutrients**: High doses of vitamin C, glutathione, and other antioxidants help combat oxidative stress and strengthen the immune system post-fast.
    - **Energy Boosters**: Nutrients like B vitamins and NAD+ help enhance mitochondrial function, boosting energy production and recovery.
    - **Hydration with Nutrients**: Combines hydration with essential vitamins and minerals to replenish what the body has used during fasting.

These nutrient drips are delivered directly into the bloodstream, bypassing the digestive system and ensuring maximum absorption. This approach focuses on feeding the cells quickly and effectively, giving the body the tools it needs to rebuild without overburdening the stomach or digestive system.

3. Nanocelle Technology for At-Home Use

For those following the NoMAD plan at home, **nanocelle technology** offers a highly effective alternative to drips. Nanocelle products are liquid nutrients designed to be absorbed under the tongue, ensuring rapid delivery to the bloodstream without relying on the digestive system.

- **How Nanocelle Products Work**:
    - These sublingual supplements are micronised into tiny particles that can easily pass through the mucous membranes in the mouth.
    - They deliver key nutrients like vitamins, minerals, and amino acids directly to the cells, providing the same benefits as a drip but in a more convenient, home-friendly format.

o Examples include nanocelle formulations of glutathione, B vitamins, and magnesium.
- **Why It's Important**:
  Nanocelle products ensure that participants can maintain and enhance their fasting experience without disrupting autophagy, making it easier for those who may not have access to in-clinic drips.

The NoMAD Philosophy: Feeding the Cell, Not the Stomach

The NoMAD Autophagy Plan is designed with the principle of "feeding the cell, not the stomach." This means providing the body with what it needs at the cellular level—such as hydration, electrolytes, and essential nutrients—without triggering digestive processes that could interrupt fasting or recovery.

- **During Fasting**: Drips and nanocelle products ensure the body stays hydrated and functional without breaking the autophagic state.
- **Post-Fasting**: These tools shift focus to rebuilding and replenishing cells, promoting recovery, and supporting long-term health.

By integrating these advanced technologies, the NoMAD Plan ensures that participants can maximise the benefits of fasting while receiving the support they need, whether they are at home or in a clinic setting. This comprehensive approach makes fasting safer, more effective, and accessible to a broader range of individuals.

How Drips, Nanocelle Technology, NAD+, and Glutathione Support the NoMAD Autophagy Plan

The NoMAD Autophagy Plan integrates advanced tools like **nutrient drips, nanocelle technology, NAD+**, and **glutathione** to enhance fasting outcomes, maintain autophagy, and support recovery. These interventions focus on "feeding the cell, not the stomach," ensuring optimal cellular function and detoxification without disrupting the fasting state.

1. Drips to Maintain Autophagy

During fasting, the body relies on cellular energy and detoxification processes, making it crucial to maintain autophagy while addressing essential needs like hydration and mineral balance.

- **Electrolyte Drips**: Provide essential minerals (e.g., magnesium, potassium) and hydration without introducing calories that might interrupt autophagy.
- **NAD+ Drips**: Support mitochondrial function and energy production. NAD+ (Nicotinamide Adenine Dinucleotide) is vital for cellular repair and energy metabolism. During fasting, NAD+ levels naturally rise, enhancing autophagy, and supplementation through drips can further optimise these processes.
- **Glutathione Drips**: A powerful antioxidant that supports the liver and aids in detoxification during fasting. It helps the body eliminate toxins released during autophagy, reducing oxidative stress and promoting cellular health.

These drips allow participants to stay in an autophagic state longer by ensuring hydration and essential nutrient support while enhancing detoxification and mitochondrial efficiency.

2. Drips to Support Recovery and Rebuilding After Fasting

After the fasting phase, the body requires targeted nutrition to rebuild and recover. This is where nutrient drips play a vital role in promoting regeneration and strengthening immunity.

- **Amino Acids**: Delivered via drips to aid in muscle repair, cellular regeneration, and overall recovery.
- **NAD+ Drips**: Boost mitochondrial function post-fast, replenishing energy stores and supporting anti-aging

pathways. NAD+ helps repair DNA and enhances cellular communication, accelerating recovery and long-term health benefits.
- **Glutathione Drips**: Post-fast, glutathione continues to play a critical role in detoxification and immune support. It helps neutralize toxins released during the fasting phase and supports the repair of tissues.
- **Energy Boosters**: B vitamins and other micronutrients are delivered to enhance energy production and metabolic recovery.

3. Nanocelle Technology for At-Home Use

For those following the NoMAD plan at home, **nanocelle technology** provides a highly effective, convenient alternative to drips. These sublingual products are designed for rapid absorption, bypassing the digestive system and delivering nutrients directly to the bloodstream.

- **NAD+ in Nanocelle Form**: This sublingual option offers the same mitochondrial and cellular energy benefits as NAD+ drips, making it ideal for home use. It supports energy metabolism and cellular repair without the need for clinical intervention.
- **Glutathione in Nanocelle Form**: A fast-absorbing option that enhances detoxification, reduces oxidative stress, and supports liver function.
- **Other Nanocelle Products**: Include magnesium, B vitamins, and amino acids, ensuring the body receives essential nutrients without breaking the fasting state.

4. The Value of Glutathione in Detoxification

Glutathione is a cornerstone of the NoMAD Autophagy Plan due to its powerful detoxification properties. As the body transitions into fasting and autophagy, it releases toxins from stored fat and cellular debris. Glutathione helps the liver process and eliminate these toxins, protecting the body from oxidative damage.

- **During Fasting**: Glutathione supports liver function and aids in neutralising free radicals released during detoxification.
- **Post-Fasting**: It continues to enhance recovery by reducing inflammation, supporting tissue repair, and improving immune resilience.

5. The Role of NAD+ in Fasting and Recovery

NAD+ is a molecule critical for energy production, DNA repair, and overall cellular health. It plays a dual role in the NoMAD plan:

- **During Fasting**: NAD+ levels naturally rise, boosting mitochondrial efficiency and facilitating autophagy. Supplementing NAD+ through drips or nanocelle products amplifies these effects, ensuring sustained energy and cellular repair.
- **Post-Fasting**: NAD+ helps rebuild energy stores, repair DNA damage, and optimise mitochondrial health, which is essential for long-term recovery and anti-aging benefits.

Post-Fasting Phase: Transitioning to Sustained Nutrition

After completing the fasting phase and 10 days of One Meal a Day (OMAD), the NoMAD Autophagy Plan transitions participants into a structured, nutrient-rich dietary approach designed to rebuild and support long-term health. This phase is carefully sequenced to maintain the benefits of autophagy while reintroducing balanced nutrition.

The Autophagy Path to Healing

Phase 1: Two Meals a Day – Predominantly Vegetables, Protein, and Healthy Fats

For the first two weeks following OMAD, participants shift to eating **two meals a day**. These meals are predominantly composed of:

- **Vegetables**: A wide variety of non-starchy vegetables, rich in fiber, antioxidants, and essential nutrients to support gut health and cellular repair.
- **Proteins**: High-quality protein sources, such as lean meats, fish, eggs, or plant-based proteins, to aid in tissue repair and regeneration.
- **Healthy Fats**: Good fats from sources like avocados, nuts, seeds, and olive oil to provide sustained energy and support hormonal balance.

This phase allows the body to gradually adjust to increased caloric intake while maintaining the metabolic and cellular benefits achieved during fasting. The focus remains on nutrient density and avoiding rapid spikes in blood sugar.

Phase 2: Reintroducing Low Glycemic Load (GL) Carbohydrates

After the initial two weeks, participants can begin to introduce **low glycemic load carbohydrates** into one-third of their meals. Examples include quinoa, sweet potatoes, or legumes. These slow-digesting carbs provide steady energy without causing significant blood sugar fluctuations.

- **Why This Approach Works:**
    o Gradual reintroduction of carbs ensures that metabolic flexibility is maintained while supporting energy demands.
    o It prevents the sudden return of high insulin spikes, helping participants sustain the benefits of autophagy and fat adaptation.

The Autophagy Path to Healing

Challenges and Observations

While this structured approach supports long-term health, many participants find it mentally and emotionally challenging, particularly during the reintroduction of carbohydrates. This is because:

- **Heightened Sensory Awareness ("Radars On")**: After fasting and OMAD, participants often report an increased sensitivity to food. This heightened awareness can make carb-rich foods feel overwhelming or overly indulgent, leading some to resist reintroducing them.
- **Preference for Protein and Fats**: Many participants naturally gravitate toward protein and fats, as they feel more satiated and aligned with their post-fasting state.

The NoMAD Plan addresses these challenges by offering personalised guidance and support, helping participants transition smoothly into a sustainable, long-term dietary approach that suits their preferences and metabolic needs.

The Broad Benefits of the NoMAD Autophagy Plan

The NoMAD Autophagy Plan is more than just a fasting protocol—it's a holistic approach to healing and rejuvenation that addresses multiple systems in the body simultaneously. Autophagy, the process of cellular cleanup and renewal, impacts **psychology, neurology, immunology, and endocrinology**, creating a ripple effect of benefits across the entire body. By reducing inflammation, autophagy supports the idea that **everything in the body works together in harmony**, rather than as isolated parts.

This plan focuses specifically on **autophagy detoxification**, a cornerstone of the **RELEASE series**, which is part of the comprehensive 6R framework for holistic health. While this book centers on cellular detox, the rest of the RELEASE series explores other critical areas of detoxification, including the brain, muscles, gut, liver, kidneys, lymphatic system, and more.

The Autophagy Path to Healing

The 6R Framework and the RELEASE Series

The NoMAD Autophagy Plan is part of the **RELEASE phase** in the broader 6R plan: Restore, Release, Repair, Rebuild, Renew, and Revisit. The RELEASE phase specifically focuses on detoxification across various systems. While this book emphasises autophagy detox, the RELEASE series covers a wide range of interconnected detox pathways:

1. Brain Detox

- **Focus**: Clearing neurotoxins, such as beta-amyloid plaques, to support cognitive health and mental clarity.
- **Mechanisms**: Enhancing autophagy in the brain and supporting the glymphatic system, which removes waste products during sleep.
- **Goal**: Reduce neuroinflammation, improve brain plasticity, and lower the risk of neurodegenerative conditions like Alzheimer's and Parkinson's.

2. Muscle Detox

- **Focus**: Addressing protein debris and metabolic waste products that accumulate in muscle tissue.
- **Mechanisms**: Autophagy clears out damaged mitochondria and proteins in muscle cells, improving muscle function and endurance.
- **Goal**: Promote muscle recovery, reduce inflammation, and support strength and flexibility.

3. Gut Detox

- **Focus**: Healing the gut lining and removing toxins that can contribute to systemic inflammation.
- **Mechanisms**: Enhancing autophagy in gut cells, improving microbiome balance, and supporting the gut-brain axis.

- **Goal**: Strengthen the gut barrier, reduce leaky gut syndrome, and improve digestion and immune health.

4. Liver and Kidney Detox

- **Focus**: Supporting the body's primary detoxification organs to process and eliminate toxins efficiently.
- **Mechanisms**: Optimising liver enzyme activity, enhancing kidney filtration, and reducing oxidative stress.
- **Goal**: Promote toxin elimination, improve metabolic health, and reduce inflammation.

5. Lymphatic Detox

- **Focus**: Enhancing the drainage of toxins and cellular waste from the lymphatic system.
- **Mechanisms**: Stimulating lymph flow through movement, hydration, and targeted therapies.
- **Goal**: Reduce swelling, improve immune function, and support systemic detoxification.

6. Heavy Metal Detox

- **Focus**: Removing environmental toxins such as mercury, lead, and arsenic from the body.
- **Mechanisms**: Using chelation strategies, glutathione supplementation, and dietary interventions to bind and eliminate heavy metals.
- **Goal**: Reduce oxidative stress and inflammation caused by heavy metal exposure, improving cellular function.

All Systems Working Together

The interconnected nature of the body means that detoxing one area often benefits others. For example:

- Clearing toxins from the **brain** enhances focus and emotional stability, which positively impacts the gut-brain axis.
- Detoxifying **muscle tissue** improves circulation, which supports the lymphatic system.
- Healing the **gut** reduces systemic inflammation, benefiting the immune system and hormonal balance.

Autophagy is a foundational tool in this process, facilitating detoxification at the cellular level while the rest of the RELEASE series targets specific systems to ensure holistic renewal.

The NoMAD Philosophy: A Unified Approach to Health

The NoMAD Autophagy Plan demonstrates how detoxification across multiple systems works synergistically to optimise overall health. Autophagy detox is the starting point, clearing cellular debris and reducing inflammation. The rest of the RELEASE series builds on this foundation by addressing the brain, muscles, gut, liver, kidneys, lymphatic system, and environmental toxins. Together, these tools provide a comprehensive strategy for achieving long-term vitality and balance.

The Autophagy Path to Healing

Summary: The NoMAD Autophagy Plan: Detoxification and Holistic Renewal

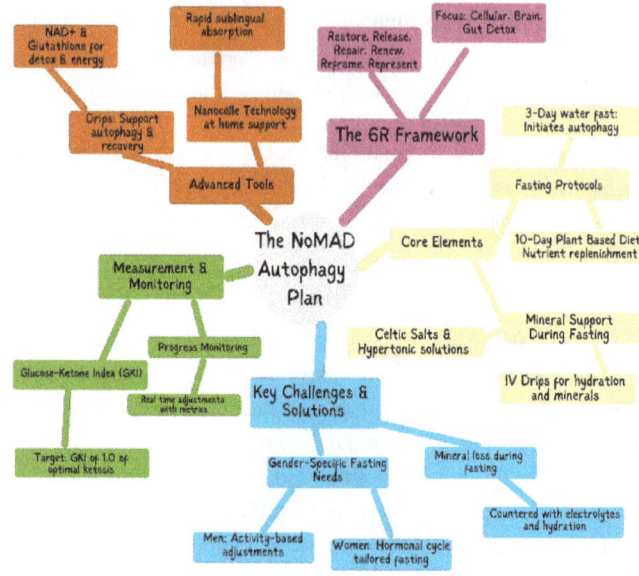

Chapter 7 Real-Life Success Stories

This chapter is not just about sharing stories—it is about real people, real struggles, and real triumphs. These success stories demonstrate the transformative power of the NoMAD Plan, showing that change is possible, no matter how long you have been stuck in unhealthy patterns. Let these stories inspire you, educate you, and remind you that you, too, can reclaim your health, vitality, and well-being.

7.1 Case Study One: Weight Loss and Energy

Sarah's Journey: Reclaiming Health and Vitality

Sarah, a 45-year-old mother of two, had been trapped in a cycle of diets, exercise plans, and constant disappointment for over twenty years. She tried everything—low-carb diets, calorie counting, intense workout regimens—but nothing seemed to stick. The weight always came back, bringing with it a deep sense of failure and fatigue. Sarah's energy levels were at an all-time low, and she found herself running on fumes just to get through each day.

When Sarah heard about the NoMAD Plan, she was understandably sceptical. "How could something as simple as fasting and mindful eating make a difference?" she wondered. But Sarah decided she had nothing to lose. She committed herself to the plan, cautiously hopeful that this time, things might be different.

To her surprise, within the first week, Sarah began to notice subtle yet profound changes. Her energy levels started to rise— no longer did she wake up feeling drained. It was as if her body had been a clogged engine, and the NoMAD Plan was the oil change it desperately needed. She started to sleep better, waking up feeling refreshed and ready to take on the day. Most remarkably, she noticed that the constant hunger and cravings that had plagued her for years began to diminish. It was as if her body was finally getting the message: you are being nourished, you are okay.

Over the next few months, Sarah's transformation continued. She lost an astonishing 76 kilos (167 pounds). But the weight loss was just the beginning. Sarah's body was finally functioning the way it was supposed to—her cells were clean, efficient, and fueled by the right kind of energy: healthy fats. It was not just about eating less; it was about nourishing her body in a way that allowed it to heal from the inside out.

Sarah's story is more than a weight-loss success; it is a journey of reclaiming life. Imagine carrying a heavy backpack for years, filled with stones of exhaustion, frustration, and hopelessness. The NoMAD Plan helped Sarah gradually remove those stones, one by one, until she could stand tall, unburdened, and full of vitality. Her journey teaches us that when you align your eating habits with your body's natural rhythms, incredible things can happen.

7.2 Case Study Two: Overcoming ADHD and Insomnia

John's Journey: Finding Focus and Peace

John was a 20-year-old college student whose life was spiralling out of control. His mind was a constant whirlwind—racing thoughts, endless distractions, and a chronic inability to focus. He had been diagnosed with Attention deficit hyperactivity disorder (ADHD), and on top of that, he suffered from insomnia. Every night felt like a battle, tossing and turning, his mind unable to shut down. Traditional treatments had not worked for him; he felt like he was losing hope.

When John stumbled upon the NoMAD Plan, he was intrigued but doubtful. How could something as straightforward as simplifying his diet and introducing fasting make a difference in his complex mental health struggles? Nevertheless, John decided to give it a try. He began with the basics: a structured fasting plan combined with clean, mindful eating.

Within weeks, John noticed a shift. It was subtle at first, like the quieting of a storm. His thoughts became clearer, more focused. For the first time in years, he could sit down to study without being overwhelmed by distractions. His energy levels became more consistent throughout the day, no longer subject to the peaks and valleys that had previously dictated his mood.

Most significantly, John's sleep improved. The racing thoughts that had kept him awake at night began to calm. It was as if his mind had been a jumbled radio, constantly changing stations, and the NoMAD Plan helped him tune into a single, precise frequency. John started to wake up feeling rested, his mind sharp and ready to take on the day.

John's story is a powerful testament to the connection between our physical and mental health. Just as a car runs better on the right fuel, our minds function best when our bodies are properly nourished and rested. The NoMAD Plan gave John the tools to find balance in his life—improving his academic performance, his relationships, and most importantly, his overall quality of life. His journey teaches us that sometimes, the answers to our most persistent struggles can be found in the simplest of changes.

7.3 Case Study Three: Reversing Early-Onset Diabetes

Emily's Journey: Turning Back the Clock on Chronic Disease

Emily was a 62-year-old businesswoman who had always been in control of her life—until she was diagnosed with early-onset diabetes. The news hit her hard. She was determined to avoid a lifetime of medication, but she did not know where to start. Her doctor, recognising Emily's resolve, suggested the NoMAD Plan as a natural way to manage her condition.

Emily was cautious but motivated. She had seen the impact of diabetes on her friends and family—medications, constant blood sugar monitoring, and the ever-present fear of complications. She was determined to take a different path. Emily embraced the NoMAD Plan, focusing on the principles of fasting and mindful eating.

The results were nothing short of miraculous. Within a few months, Emily's blood sugar levels stabilised. Her doctor was amazed—she had effectively reversed her diabetes without the need for medication. But the benefits did not stop there. Emily felt a new sense of mental clarity as if a fog had lifted. Her energy levels soared, and she found herself more productive and focused at work than she had been in years.

Emily's journey is a beacon of hope for anyone struggling with chronic conditions. It is a reminder that our bodies have an incredible capacity to heal when given the right conditions. Imagine your body as a garden; when you remove the weeds of poor diet and lifestyle habits and nourish the soil with the proper nutrients, it is incredible how quickly it can flourish. Emily's story teaches us that it is never too late to take control of your health and that with the right plan, you can turn back the clock on chronic disease.

7.4 Case Study Four: Managing Cravings and Stress

David's Journey: Regaining Control and Finding Balance

David was a successful businessman, the kind of person who thrived under pressure. But behind the facade of success, he was constantly battling cravings and stress. Despite his best efforts to eat healthily, David found himself reaching for junk food during late-night work sessions. His weight was creeping up, and his health was beginning to suffer. Worse still, the stress of his job was taking a toll on his mental and emotional well- being.

The Autophagy Path to Healing

When David heard about the NoMAD Plan, he was intrigued by its simplicity. Could simplifying his diet and lifestyle really help him manage his cravings and stress? David decided to find out. He started with the basics: reducing processed foods, incorporating fasting, and focusing on mindful eating.

The changes were gradual but profound. David began to notice that his cravings for unhealthy foods were diminishing. It was as if his body was learning to ask for what it truly needed rather than what it had been conditioned to want. As his diet became simpler and more aligned with his body's needs, David found that he was better able to manage the stress in his life. The late-night junk food binges were replaced by healthy snacks, and his weight started to come down.

But the most significant change was in how David felt. He began to sleep better, wake up feeling refreshed, and approach his work with a new sense of calm and focus. It was as if the NoMAD Plan had helped him regain control—not just over his eating habits, but over his life as a whole.

David's story is a powerful reminder that managing stress and cravings is not about willpower alone; it is about giving your body the right tools to function optimally. Imagine your body as a ship; when the seas are rough, it is easy to be tossed around by cravings and stress. But with the NoMAD Plan as your compass, you can steer your ship through the storm, finding calm waters and a renewed sense of direction. David's journey teaches us that by simplifying our diet and lifestyle, we can regain control, reduce stress, and live a more balanced, fulfilling life.

These expanded stories not only highlight the profound impact of the NoMAD Plan but also serve as an educational and inspirational guide for anyone looking to improve their health and well-being. They demonstrate that with the right approach, even the most daunting challenges can be overcome, and a brighter, healthier future is within reach.

The Autophagy Path to Healing

Summary: Real-Life Success Stories

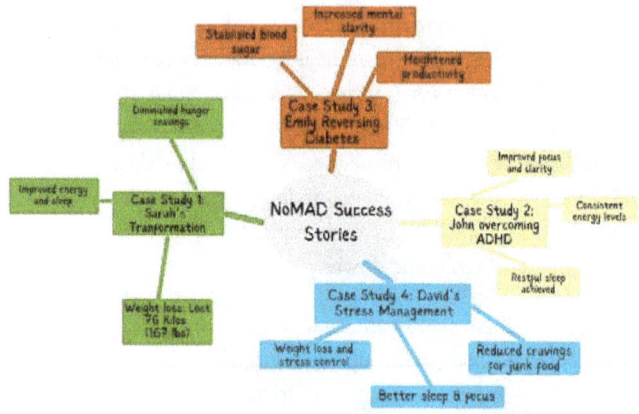

Chapter 8: Decompressing the Body Through Autophagy

Imagine your body as a finely tuned engine that has been running non-stop for years, accumulating wear and tear, clogged filters, and unaddressed inefficiencies. Autophagy—the body's natural cleaning mechanism—is like a scheduled overhaul, clearing out the old, damaged parts and making room for optimal performance.

Autophagy, derived from the Greek words "auto" (self) and "phagy" (eating), is the body's way of self-maintenance. It is the process where cells identify damaged components, break them down, recycle the materials, and create new, healthier cellular structures. Beyond a simple clean-up, autophagy decompresses the body at every level—from individual cells to entire organ systems—unlocking its ability to heal, regenerate, and thrive.

The **NoMAD Autophagy Plan** is designed to activate and maximise this profound process, targeting key organs like the gut, liver, kidneys, brain, muscles, spleen, and heart. This chapter explores how autophagy decompresses these systems, its impact on chronic diseases, and its emerging role in cancer care at pioneering centers around the world.

8.1 Autophagy: The Body's Ultimate Reset

Autophagy is more than a biological process; it is the cornerstone of cellular health. Under normal conditions, cells accumulate waste—damaged proteins, organelles, and dysfunctional mitochondria. Left unchecked, this cellular "junk" clogs systems and contributes to disease.

The Autophagy Path to Healing

When autophagy is activated, the body:

- **Cleans House**: Cellular waste is broken down and removed, clearing pathways for normal function.
- **Recycles Resources**: Broken components are repurposed into raw materials for new cellular structures.
- **Reduces Inflammation**: By eliminating debris, autophagy calms inflammatory signals that drive chronic disease.

The process unfolds systemically, decompressing organs and systems, much like deflating an overburdened balloon to restore its optimal shape and function.

8.2 The Gut: Where Health Begins

The gut is often called the body's second brain, a hub where nutrient absorption, immune defense, and even mood regulation begins. However, modern diets, stress, and toxins overload this system, leading to conditions like leaky gut, irritable bowel syndrome (IBS), and autoimmune diseases.

Autophagy in the Gut:

- **Repairs the Gut Lining**: Autophagy restores the intestinal barrier, preventing harmful toxins and pathogens from entering the bloodstream.
- **Eliminates Dysfunctional Microbes**: By targeting and recycling damaged gut bacteria, autophagy creates a balanced microbiome.
- **Improves Nutrient Absorption**: A clean, functioning gut absorbs vitamins, minerals, and amino acids more efficiently.

Studies show that enhancing autophagy in the gut can alleviate conditions like Crohn's disease and ulcerative colitis by reducing inflammation and repairing damage.

The Autophagy Path to Healing

83 The Liver: The Body's Detox Hub

The liver processes toxins, hormones, and metabolic waste, but when overburdened, it can become sluggish and inefficient. Fatty liver disease, for example, is a condition where excess fat impairs liver function, often due to poor diet or alcohol consumption.

Autophagy in the Liver:

- **Clears Fat Deposits**: Fasting-induced autophagy removes fat from liver cells, reversing conditions like non-alcoholic fatty liver disease (NAFLD).
- **Rejuvenates Detox Pathways**: Damaged liver cells are replaced, enhancing the organ's ability to filter toxins.
- **Optimises Hormonal Balance**: The liver's role in metabolising hormones is improved, addressing issues like estrogen dominance.

Fasting protocols in the **NoMAD Autophagy Plan** have been shown to regenerate liver function, supporting everything from digestion to hormonal health.

8.4 The Kidneys: Precision Filters

The kidneys filter around 50 gallons of blood daily, maintaining electrolyte balance and removing waste. Over time, toxins and debris accumulate, leading to chronic kidney disease (CKD) and reduced efficiency.

Autophagy in the Kidneys:

- **Eliminates Damaged Cells**: Dysfunctional cells in the renal tubules are cleared, improving filtration.
- **Reduces Fibrosis**: Autophagy prevents scarring in kidney tissues, a key driver of CKD progression.
- **Balances Electrolytes**: Restoring cellular health ensures proper sodium, potassium, and calcium levels.

Emerging research highlights fasting-induced autophagy as a non-invasive approach to slow kidney disease and enhance overall renal function.

8.5 The Brain: A Clearer Mind

The brain, a high-energy organ, is highly sensitive to the accumulation of waste products like amyloid plaques, which are linked to Alzheimer's disease. Neuronal autophagy helps protect brain cells and supports cognitive health.

Autophagy in the Brain:

- **Removes Plaques and Tangles**: Proteins like beta-amyloid are cleared, reducing the risk of neurodegenerative diseases.
- **Recycles Mitochondria**: Damaged mitochondria are replaced, boosting energy and reducing oxidative stress.
- **Supports Synaptic Plasticity**: The brain's ability to form and adapt connections is enhanced.

Ketones produced during fasting provide a clean and efficient energy source, further supporting brain function. Neurological conditions like Parkinson's and Alzheimer's benefit from therapies that enhance autophagy, showing promise in slowing disease progression.

8.6 The Heart: Resilient and Efficient

The heart, a tirelessly working muscle, depends on efficient energy use and minimal oxidative stress. Autophagy ensures cardiac cells remain functional and adaptable.

Autophagy in the Heart:

- **Prevents Mitochondrial Dysfunction**: Ensures the heart's energy systems are robust and efficient.
- **Reduces Cardiac Inflammation**: Removes damaged proteins and organelles that contribute to heart failure.

- **Strengthens Muscle Cells**: Rejuvenates cardiomyocytes (heart muscle cells), improving resilience.

Research shows that fasting-induced autophagy reduces the risk of heart disease, including arrhythmias and atherosclerosis, by improving cardiac efficiency.

8.7 Muscles: Strength and Recovery

Muscle tissue is constantly breaking down and rebuilding, especially with physical activity or aging. Sarcopenia, the age-related loss of muscle mass, is linked to reduced autophagy.

Autophagy in Muscles:

- **Clears Damaged Fibers**: Dysfunctional components are removed, promoting recovery and growth.
- **Enhances Energy Use**: Mitochondrial recycling optimises ATP (energy) production.
- **Prevents Muscle Wasting**: Supports muscle preservation in conditions like aging or prolonged inactivity.

Athletes use fasting to activate autophagy, improving recovery times and reducing the risk of injury.

8.8 The Spleen: Immunity's Gatekeeper

The spleen filters blood and recycles old red blood cells, playing a vital role in immune defense. Autophagy enhances spleen function by:

- **Rejuvenating Immune Cells**: Dysfunctional white blood cells are replaced, strengthening immunity.
- **Reducing Systemic Inflammation**: Removes cellular debris that triggers chronic immune responses.
- **Enhancing Infection Response**: A well-functioning spleen can better detect and fight pathogens.

This immune recalibration is critical for preventing autoimmune conditions and improving overall resilience.

8.9 Cancer and Autophagy: Supporting Modern Therapies

Autophagy's role in cancer is complex. While impaired autophagy can enable cancer growth, controlled activation helps:

- **Prevent Cancer Formation**: Removes precancerous cells and DNA damage before mutations can occur.
- **Enhance Treatment Effectiveness**: Autophagy makes cancer cells more sensitive to chemotherapy and radiation.
- **Suppress Tumor Growth**: Limits nutrient availability to cancer cells.

Centers Using Autophagy in Cancer Care:

- **Fasting Mimicking Diet (FMD)**: Developed by Dr. Valter Longo, this protocol is used to improve chemotherapy outcomes.
- **Germany's Integrative Oncology Clinics**: Combine fasting with hyperthermia and natural therapies to support cancer patients.
- **Japan's Research on Spermidine**: Studies show that autophagy-enhancing compounds like spermidine may prevent cancer progression.

These innovative approaches demonstrate the potential of autophagy as an adjunct to conventional cancer treatments.

8.10 Systemic Benefits: Healing from Within

The beauty of autophagy lies in its ability to create systemic harmony:

- **Chronic Diseases**: Conditions like diabetes, arthritis, and hypertension improve with reduced inflammation and cellular renewal.

- **Longevity**: Autophagy slows aging by preventing the accumulation of damaged components.
- **Emotional and Mental Health:** By decompressing the gut-brain axis, autophagy enhances emotional stability and cognitive function.

This all-encompassing impact positions autophagy as a cornerstone of health optimisation.

Conclusion

The **NoMAD Autophagy Plan** is a roadmap to decompression and renewal at every level of the body. By activating this natural process, the plan helps heal organs, restore balance, and optimise health. From chronic disease management to cutting-edge cancer care, autophagy is unlocking new possibilities for vitality and longevity.

The Autophagy Path to Healing

Summary: Decompressing the Body Through Autophagy

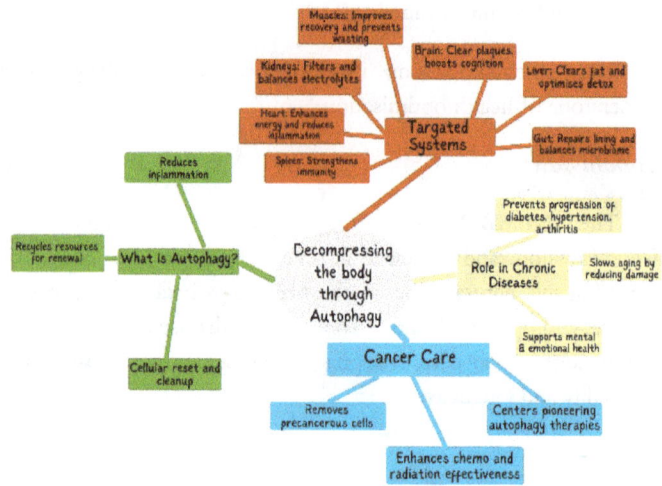

Chapter 9: Final Thoughts

Having treated patients over the last several years with various plans, I am continually awed by the resilience and adaptability of the human body. We are made up of **30 trillion cells**, coexisting harmoniously with an estimated **38 trillion bacteria**, a delicate balance that enables us to face countless threats each day. Every breath, every movement, every thought is supported by this intricate system that strives to keep us alive and thriving despite an ever-changing environment.

Yet, this resilience is not without limits. Our modern world changes faster than our biology can adapt. Our cells, still optimised for the savanna, now face industrialised food, polluted air, artificial light, chronic stress, and sedentary living. This disconnect drives **chronic inflammation, metabolic imbalances, and the accumulation of cellular debris**, which in turn fuels a cycle of decline. Over time, these unchecked processes manifest as the chronic diseases that now dominate global health.

The solution lies not in ever-escalating pharmaceutical interventions but in turning inward—to the innate intelligence of our cells. By focusing on **clearing the "toxic larders"** within us and supporting our body's natural processes, we can break free from the cycle of disease. The body holds the power to heal and regenerate; we simply need to give it the conditions to do so.

The **6 Rs—Restore, Release, Repair, Renew, Reframe, and Represent**—provide a roadmap for this transformation:

To re-summarise :

- **Restore:** Replenish the foundational nutrients and systems that support health.
- **Release:** Detoxify and remove cellular debris to reduce inflammation and clear blocked pathways.

- **Repair**: Activate the body's healing mechanisms to mend damage at the cellular and systemic level.
- **Renew**: Rejuvenate tissues, organs, and energy systems to restore vitality.
- **Reframe**: Shift perspectives and behaviors, aligning with practices that sustain long-term health.
- **Represent**: Take ownership of your health and inspire those around you to do the same.

The final R, Represent, is particularly significant. It is a call to action—a reminder that we have the power not only to transform ourselves but to lead by example, empowering others to take control of their own health. By representing the possibilities of healing and vitality, we create a ripple effect that extends to our families, communities, and beyond.

This journey is about progress to perfection – the constant pursuit of better 'Our bodies are incredibly capable when given the chance to thrive. Imagine a world where individuals clean their internal "larders" regularly, where inflammation is replaced with balance, and where chronic diseases are no longer inevitable.

The human form is a marvel of biological ingenuity, and its ability to heal continues to astonish me. The greatest medicine already lies within us. By aligning with the principles of the **6 Rs**, we not only take steps toward our own renewal but inspire a future where health is reclaimed—not through dependency on external solutions but by trusting the profound wisdom of our cells. Together, let us restore, release, repair, renew, reframe, and represent a new standard for health and vitality.

The Autophagy Path to Healing

Summary: Final Thoughts

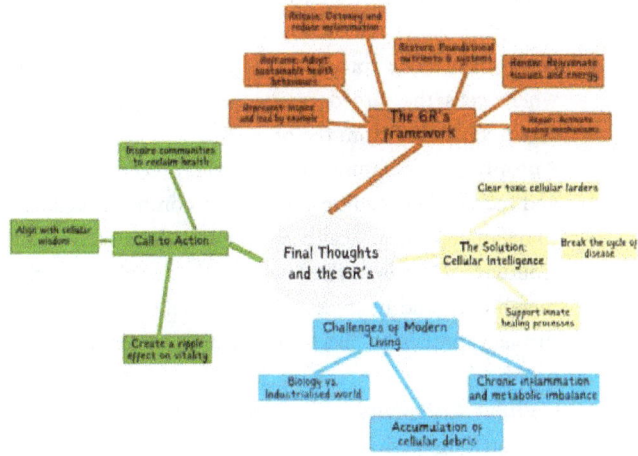

Glossary

1. **Antioxidants**: Molecules that neutralise free radicals, reducing oxidative stress and protecting cells from damage.
2. **Apoptosis**: The process of programmed cell death, where cells that are damaged or no longer needed are safely eliminated from the body.
3. **Autophagy**: A natural process where cells break down and recycle damaged or unnecessary components, helping to maintain cellular health and prevent diseases.
4. **Cellular Debris**: The waste products of cellular metabolism and dead cells, which need to be cleared out through processes like autophagy to maintain cellular health.
5. **Cellular Longevity**: The ability of cells to function well over a long period, contributing to overall health and lifespan.
6. **Cellular Metabolism**: The set of chemical reactions in cells that convert food into energy and build cellular components, essential for maintaining life.
7. **Cellular Restoration**: The process by which cells repair and renew themselves, crucial for maintaining overall health and preventing disease.
8. **Chronic Inflammation**: A prolonged inflammatory response that can damage tissues and contribute to various diseases, including heart disease and cancer.
9. **Circadian Rhythm**: The natural, internal process that regulates the sleep-wake cycle and repeats roughly every 24 hours, influencing various bodily functions.
10. **Detoxification**: The body's natural process of removing toxins and waste products, which can be enhanced through diet, fasting, and lifestyle adjustments.
11. **Epigenetic Modifications**: Changes to gene expression that do not involve alterations to the DNA sequence, often influenced by environmental factors and lifestyle.

The Autophagy Path to Healing

12. **Epigenetics**: The study of how behaviours and environment can cause changes that affect the way your genes work, which can influence health outcomes.
13. **Fatty Acids**: The building blocks of fats in the body, which are used for energy, building cell membranes, and producing signalling molecules.
14. **Glucose Ketone Index (GKI)**: A measure of the ratio of glucose to ketones in the blood, used to assess metabolic health and the effectiveness of fasting or ketogenic diets.
15. **Glycogen**: A form of stored glucose, primarily found in the liver and muscles, which is used by the body for energy.
16. **Homeostasis**: The state of steady internal conditions maintained by living organisms, essential for health and functioning.
17. **Hormetic Stress**: A beneficial form of stress that activates adaptive responses in the body, such as autophagy, to improve health and resilience.
18. **HPA Axis**: The hypothalamus-pituitary-adrenal axis, a complex set of interactions among these glands that controls stress response, digestion, immune system, and energy usage.
19. **Hypertonic Saline**: A solution with a higher concentration of salt than normal body fluids, sometimes used to restore electrolyte balance during fasting or illness.
20. **Inflammation**: The body's response to injury or infection, which can become harmful when it is chronic, leading to various diseases.
21. **Intermittent Fasting**: An eating pattern that alternates between periods of eating and fasting, which can trigger autophagy and improve metabolic health.
22. **Ketones**: Chemicals produced by the liver when fat is broken down for energy, particularly during fasting or carbohydrate- restricted diets.
23. **Ketosis**: A metabolic state where the body burns fat for energy instead of carbohydrates, often triggered by fasting or low-carbohydrate diets.

24. **Lipid Profile**: A blood test that measures fats and fatty substances in the body, such as cholesterol and triglycerides, to assess the risk of heart disease.
25. **Metabolic Health**: The state of metabolic processes in the body, including blood sugar regulation, fat metabolism, and energy production, which are critical for overall health.
26. **Mindful Adaptation**: Making thoughtful changes to lifestyle and diet that align with health goals, promoting balance and reducing stress.
27. **Mindful Eating**: Paying attention to what and how you eat, focusing on the experience rather than eating on autopilot, which supports healthier eating habits.
28. **mTOR Pathway**: A cellular signalling pathway that regulates growth, metabolism, and autophagy. It is often active when nutrients are plentiful and suppressed during fasting.
29. **NoMAD Approach**: A health plan focused on Nutritional Optimisation, Mindful Adaptation, and Detoxification, designed to support autophagy and overall well-being.
30. **Nutrient Scarcity**: A state in which the body has limited access to nutrients, often triggering autophagy as the body seeks to conserve resources.
31. **Nutrient-Dense Foods**: Foods that are high in vitamins, minerals, and other essential nutrients but low in calories, promoting health without excess energy intake.
32. **Nutritional Optimisation**: Eating in a way that supports the body's needs without excess, often by focusing on nutrient- dense, whole foods.
33. **Oxidative Stress**: A condition characterised by excessive free radicals in the body, which can damage cells and contribute to ageing and disease.
34. **Sirtuins**: A family of proteins that play a role in cellular health, ageing, and metabolism, often activated by fasting or calorie restriction.

35. **Visceral Fat**: Fat that is stored within the abdominal cavity and surrounds internal organs, which is linked to higher risks of metabolic diseases.

References

1. Chen, G., et al. (2014). Autophagy, Inflammation, and Immune Responses. *Autophagy*, 10(2), 238-241.
2. Chen, G., et al. (2017). Autophagy: A Double-Edged Sword for Neuronal Survival After Ischemic Stroke. *Neuroscience Bulletin*, 33(5), 497-508.
3. Choi, A. M., Ryter, S. W., & Levine, B. (2013). Autophagy in Human Health and Disease. *New England Journal of Medicine*, 368(19), 1845-1846.
4. Ciechanover, A. (2005). Intracellular Protein Degradation: From a Vague Idea Through the Lysosome and the Ubiquitin- Proteasome System and onto Human Diseases and Drug Targeting. *EMBO Journal*, 24(6), 1307-1320.
5. Cuervo, A. M., & Macian, F. (2014). Autophagy, Nutrition and Immunity. *Nature Reviews Immunology*, 14(6), 492-505.
6. Deretic, V. (2012). Autophagy: An Emerging Immunological Paradigm. *Journal of Immunology*, 189(1), 15-20.
7. Deretic, V., et al. (2013). Autophagy in Infection, Inflammation and Immunity. *Nature Reviews Immunology*, 13(10), 722-737.
8. Eskelinen, E. L., & Saftig, P. (2009). Autophagy: A Lysosomal Degradation Pathway with a Central Role in Health and Disease. *Biochimica et Biophysica Acta (BBA) - Molecular Cell Research*, 1793(4), 664-673.
9. Galluzzi, L., et al. (2015). Autophagy in Malignant Transformation and Cancer Progression. *EMBO Journal*, 34(7), 856-880.
10. Galluzzi, L., et al. (2017). Autophagy in Health and Disease: A Report from the 1st International Forum on Autophagy. *Autophagy*, 13(2), 224-247.
11. Gatica, D., Lahiri, V., & Klionsky, D. J. (2018). Cargo Recognition and Degradation by Selective Autophagy. *Nature Cell Biology*, 20(3), 233-242.

12. Glick, D., Barth, S., & Macleod, K. F. (2010). Autophagy: Cellular and Molecular Mechanisms. *Journal of Pathology*, 221(1), 3-12.
13. Grahame Hardie, D. (2014). AMP-activated protein kinase: a key regulator of energy balance with many roles in human disease. Journal of Internal Medicine.276 (6) 543-559.
14. Green, D. R., & Levine, B. (2014). To Be or Not to Be? How Selective Autophagy and Cell Death Govern Cell Fate. *Cell*, 157(1), 65-75.
15. Guo, H., & Chien, S. (2020). Role of Autophagy in Cellular Response to Mechanical Stretching. *Journal of Cellular Physiology*, 235(1), 1-9.
16. Hansen, M., & Rubinsztein, D. C. (2013). Autophagy as a Therapeutic Target for Age-Related Diseases. *Trends in Pharmacological Sciences*, 34(8), 483-490.
17. He, C., & Klionsky, D. J. (2009). Regulation Mechanisms and Signaling Pathways of Autophagy. *Annual Review of Genetics*, 43, 67-93.
18. He, C., & Levine, B. (2010). The Beclin 1 Interactome. *Current Opinion in Cell Biology*, 22(2), 140-149.
19. Jiang, P., & Mizushima, N. (2014). Autophagy and Human Diseases. *Cell Research*, 24(1), 69-79.
20. Kaur, J., & Debnath, J. (2015). Autophagy at the Crossroads of Catabolism and Anabolism. *Nature Reviews Molecular Cell Biology*, 16(8), 461-472.
21. Kirkin, V., & Rogov, V. V. (2019). A Diversity of Selective Autophagy Receptors Determines the Specificity of the Autophagy Pathway. *Molecular Cell*, 76(2), 268-285.
22. Klionsky, D. J. (2007). Autophagy: From Phenomenology to Molecular Understanding in Less Than a Decade. *Nature Reviews Molecular Cell Biology*, 8(11), 931-937.
23. Klionsky, D. J. (2008). Autophagy Revisited: A Conversation with Christian de Duve. *Autophagy*, 4(6), 740-743.

24. Klionsky, D. J., & Emr, S. D. (2000). Autophagy as a Regulated Pathway of Cellular Degradation. *Science*, 290(5497), 1717-1721.
25. Klionsky, D. J., et al. (2016). Guidelines for the Use and Interpretation of Assays for Monitoring Autophagy. *Autophagy*, 12(1), 1-222.
26. Kroemer, G., & Levine, B. (2008). Autophagy: An Evolutionarily Conserved Process Underlying Cell Death and Survival. *Annual Review of Biochemistry*, 77, 695-726.
27. Kroemer, G., & White, E. (2010). Autophagy for the Avoidance of Neurodegeneration. *Genes & Development*, 24(9), 1181-1183.
28. Levine, B., & Kroemer, G. (2008). Autophagy in the Pathogenesis of Disease. *Cell*, 132(1), 27-42.
29. Levine, B., & Packer, M. (2015). Autophagy in the Pathogenesis of Heart Disease. *Circulation Research*, 116(3), 456-467.
30. Lilja S, Stoll C, Krammer U, Hippe B, Duszka K, Debebe T, Höfinger I, König J, Pointner A, Haslberger A. Five Days Periodic Fasting Elevates Levels of Longevity Related *Christensenella* and Sirtuin Expression in Humans. *International Journal of Molecular Sciences*. 2021; 22(5):2331.
31. Madeo, F., et al. (2015). Caloric Restriction Mimetics: Towards a Molecular Definition. *Nature Reviews Drug Discovery*, 13(10), 727-740.
32. Marino, G., et al. (2014). Autophagy: An Essential Intracellular Mechanism to Survive Stress and Maintain Health. *Journal of Pathology*, 231(1), 10-24.
33. Mizushima, N. (2007). Autophagy: Process and Function. *Genes & Development*, 21(22), 2861-2873.
34. Mizushima, N. (2018). A Brief History of Autophagy from Cell Biology to Physiology and Disease. *Nature Cell Biology*, 20(5), 521-527.
35. Mizushima, N., & Komatsu, M. (2011). Autophagy: Renovation of Cells and Tissues. *Cell*, 147(4), 728-741.

36. Mizushima, N., & Levine, B. (2010). Autophagy in Mammalian Development and Differentiation. *Nature Cell Biology*, 12(9), 823-830.
37. Mizushima, N., & Yoshimori, T. (2004). How to Interpret LC3 Immunoblotting. *Autophagy*, 1(1), 72-76.
38. Nixon, R. A. (2013). The Role of Autophagy in Neurodegenerative Disease. *Nature Medicine*, 19(8), 983-997.
39. Rubinsztein, D. C., et al. (2011). Autophagy and Aging. *Cell*, 146(5), 682-695.
40. Saftig, P., & Klumperman, J. (2009). Lysosome Biogenesis and Lysosomal Membrane Proteins: Trafficking Meets Function. *Nature Reviews Molecular Cell Biology*, 10(9), 623-635.
41. Schneider, J. L., & Cuervo, A. M. (2014). Autophagy and Human Disease: Emerging Themes. *Current Opinion in Genetics & Development*, 26, 16-23.
42. Singh, R., & Cuervo, A. M. (2011). Autophagy in the Cellular Energetic Balance. *Cell Metabolism*, 13(5), 495-504.
43. Sinha, R. A., et al. (2015). Autophagy as a Therapeutic Target for Hepatic Disorders. *Hepatology*, 61(1), 191-202.
44. Thorburn, A. (2014). Autophagy and Disease. *Journal of Biological Chemistry*, 289(37), 26155-26161.
45. Yang, Z., & Klionsky, D. J. (2010). Eaten Alive: A History of Macroautophagy. *Nature Cell Biology*, 12(9), 814-822.
46. Yoshimori, T., & Noda, T. (2008). Toward Understanding Autophagy: Lessons from Yeast. *Biochemical Society Transactions*, 36(1), 12-16.
47. Yoshinori Ohsumi. (2016). Autophagy from Beginning to End: How the Random Yeast Mutation Became a Nobel Discovery. *Cell*, 167(3), 725-728.

www.ingramcontent.com/pod-product-compliance
Lightning Source LLC
Chambersburg PA
CBHW070033040426
42333CB00040B/1638